T0134926

Sustainable Food Drying Techniques in Developing Countries: Prospects and Challenges

Mahadi Hasan Masud • Azharul Karim
Anan Ashrabi Ananno • Asif Ahmed

Sustainable Food Drying Techniques in Developing Countries: Prospects and Challenges

 Springer

Mahadi Hasan Masud
School of Engineering
RMIT University
Bundoora Campus
Melbourne, VIC, Australia

Department of Mechanical Engineering
Rajshahi University of Engineering
and Technology
Rajshahi, Bangladesh

Anan Ashrabi Ananno
Department of Mechanical Engineering
Rajshahi University of Engineering
and Technology
Rajshahi, Bangladesh

Azharul Karim
Science and Engineering Faculty
Queensland University of Technology
Brisbane, QLD, Australia

Asif Ahmed
Department of Mechanical Engineering
Rajshahi University of Engineering
and Technology
Rajshahi, Bangladesh

ISBN 978-3-030-42478-7   ISBN 978-3-030-42476-3   (eBook)
https://doi.org/10.1007/978-3-030-42476-3

This Springer imprint is published by the registered company Springer Nature Switzerland AG
The registered company address is: Gewerbestrasse 11, 6330 Cham, Switzerland

This book is dedicated to our beloved family members and to the people of the developing countries.

Preface

According to the World Food Programme, an estimated 821 million people are vulnerable to acute hunger. Most of them live in the developing and least developed countries. The world hunger crisis is exacerbated by the excessive amount of food waste (1.3 billion tons each year). Although drying techniques have substantiated itself as an effective means of reducing food waste, they are very costly, energy-intensive, and time-consuming process. The developing countries are not only economically vulnerable but also have limited energy resources to undertake necessary measures to reduce food waste and, therefore, cannot adopt expensive industrial drying techniques for long-term food preservation. This book, therefore, aims to impart knowledge on energy scenarios of the developing countries, availability of possible energy sources for drying, and, based on these, proposes some renewable energy-based sustainable drying techniques for these countries. It highlights the urgency of using renewable energy-based food drying techniques. It also discusses essential food properties that appeal to the general user and the procedure of retaining those quality aspects using an appropriate drying method and presents drying technologies that are currently practiced in various developing countries, as well as their pros and cons. This book proposes ten sustainable drying techniques for developing countries, which can potentially reduce annual food waste generation. Moreover, this book extensively examines the working principle, efficiency, effectivity, and dried food quality of the proposed drying techniques. Each developing country has a unique energy scenario and geographical features. Therefore, an in-depth analytical comparison among all the proposed dryers was presented. Industrial application of new drying technologies can be challenging irrespective of a country's economic background, and therefore, this book has a dedicated chapter to analyze the challenges that may arise while adopting the proposed sustainable drying techniques. Food security is a fundamental human right. This book is written with an intention to abate the abject condition of food waste.

Melbourne, VIC, Australia
Rajshahi, Bangladesh   Mahadi Hasan Masud
Brisbane, QLD, Australia   Azharul Karim
Rajshahi, Bangladesh   Anan Ashrabi Ananno
Rajshahi, Bangladesh   Asif Ahmed

Acknowledgments

At the very beginning, we express our utmost gratitude to the Almighty Creator for His gracious help to accomplish this work. We would like to thank our families for all their inspiration, love, and unconditional support. Also, we would like to express our heartiest gratitude to many people who have supported us in our journey of writing this book.

We sincerely acknowledge the numerous scientific discussions and assistance given by our colleagues at Rajshahi University of Engineering & Technology. We would especially like to thank Associate Professor Mohammad Uzzal Hossain Joardder for his crucial scientific discussions on the drying of food materials.

We would like to thank Springer, the publisher of this book, for allowing us to share our knowledge and findings with the research community. Special thanks go to Daniel Falatko and Harithashrivarshini Somasundaram Balasubramanian, who have keenly worked with us from the inception to the completion of this project.

Contents

About the Authors

Mahadi Hasan Masud received his BSc and MSc in Mechanical Engineering from Rajshahi University of Engineering & Technology (RUET). He currently serves as a Faculty Member in Mechanical Engineering of RUET. Moreover, he is currently pursuing his PhD at RMIT University, Melbourne, Australia. His research focus is on advanced food preservation techniques, simultaneous heat and mass transfer, renewable energy resources, and biomimicry-inspired vehicle design. He authored 3 popular books published by Springer Nature, 6 book chapters, more than 20 refereed journal publications, and more than 10 international conference publications. Most of his journal articles are published in highly ranked journals. He is a regular Reviewer of several high-ranked journals of prominent publishers, including Elsevier, Springer Nature, and Taylor & Francis.

Azharul Karim is currently working as an Associate Professor in the Mechanical Engineering Discipline, Science and Engineering Faculty, Queensland University of Technology, Australia. He received his PhD degree from Melbourne University in 2007. Through his scholarly, innovative, high-quality research, he has established a national and international standing. He has authored over 194 peer-reviewed articles, including 94 high-quality journal papers, 13 peer-reviewed book chapters, and 4 books. His papers have attracted about 3100 citations with an h-index of 30. His papers have very high impact worldwide, as demonstrated by his overall field weighted citation index (FWCI) which is 2.99. He is Editor/Board Member of six reputed journals, including *Drying Technology* and *Scientific Reports*, and Supervisor of 26 past and current PhD students. He has been Keynote/Distinguished Speaker at scores of international conferences and Invited/Keynote Speaker in seminars in many reputed universities worldwide. He won multiple international awards for his outstanding contributions to multidisciplinary fields. His research is directed toward solving acute food industry problems by advanced multiscale and multiphase food drying models of cellular water using theoretical/computational and experimental methodologies. His global leadership in the fields of food drying and microwave drying is evident by the high impact of his papers, which is significantly higher than world averages. His current research areas are food drying and its multiscale and multiphase modelling, nanofluid solar thermal storage, concentrating PV-thermal collector, and lean healthcare systems.

Anan Ashrabi Ananno completed his BSc in Mechanical Engineering from Rajshahi University of Engineering & Technology (RUET). His research focus is on advanced food preservation techniques, simultaneous heat and mass transfer, renewable energy resources, waste management practices, biomimicry and data science. He has authored one book, six refereed journals, four book chapters, and three international conference papers. Most of his journal articles are published in regular issues of Elsevier, Springer, and Taylor & Francis. He is also a Regular Reviewer of several high-ranked journals of prominent publishers, including Elsevier and Springer. He has won multiple international awards in youth innovation challenges and model united nations.

Asif Ahmed completed his BSc in Mechanical Engineering from Rajshahi University of Engineering & Technology. Besides academic career, he works with voluntary social organizations. His current research areas are food drying, renewable energy utilization, and waste management. He authored couple of international peer-reviewed journals and conference publications.

List of Abbreviations

UN	United Nations
GDP	Gross Domestic Product
FAO	Food and Agricultural Organization
MDG	Millennium Development Goal
EST	Environmentally Sound Technologies
TNA	Technology Needs Assessment
UNFCC	United Nations Framework Convention on Climate Change
GHG	Greenhouse Gas
IEA	International Energy Agency
CIS	Commonwealth of Independent States
EU	European Union
BPD	Barrels Per Day
OECD	Organisation for Economic Co-operation and Development
UV	Ultraviolet
IUPAC	International Union of Pure and Applied Chemistry
JIRCAS	Japan International Research Center for Agricultural Sciences
WHO	World Health Organization
PCM	Phase Change Material
EUT	Earth's Undisturbed Temperature
SHA	Solar Heat Absorber
PA-FPSC	PCM Flat-Plate Solar Collector
FPSC	Standalone Flat-Plate Solar Collectors
EAHX	Earth to Air Heat Exchanger
WAHX	Water-Air Heat Exchanger
COP	Coefficient of Performance
SMER	Specific Moisture Extraction Rate
SEC	Specific Energy Consumption
NTNU	Norwegian Institute of Technology
WHCD	Waste Heat Convective Dryer
EGHR	Exhaust Gas Heat Recovery System
HFO	Heavy Fuel Oil

PMSD	Pulsed Microwave-Solar Dryer
IMCD	Intermittent Microwave Convective Drying
MCD	Convective Microwave Drying
MWC	Microwave Convective
CD	Convective Drying
CFC	Chlorofluorocarbons
NOx	Oxides of Nitrogen
QBtu	Quadrillion British Thermal Units
TBtu	Trillion British Thermal Units
db	Dry Basis
wb	Wet Basis

Nomenclature

Symbol	Meaning
MC_c	Critical moisture content
a_w	Water activity
$°C$	Temperature in Celsius
$\%wb$	Percent of wet bulb
Q_e	Energy (KJ)
ω_1	Humidity ratio of air at point 1 (kg water/kg dry air)
ω_2	Humidity ratio of air at point 2 (kg water/kg dry air)
dx	Smallest change of distance in x direction
q	Heat transfer (KJ)
T_p	Temperature of absorber plate (K)
T_g	Temperature of glass plate (K)
T_s	Temperature of the sky (K)
T_{in}	Inlet air temperature (K)
T_i	Temperature in insulation (K)
T_p	Temperature in absorber plate (K)
T_a	Absolute air temperature (K)
T_{wall}	Wall temperature (K)
$T_{in, EAHX}$	Pipe fluid inlet temperature (K)
$T_{Out, EAHX}$	Pipe fluid outlet temperature (K)
$T_{z, t, EAHX}$	Earth's undisturbed temperature (K)
$T_{m, EAHX}$	Soil average surface temperature at z (meter) depth (K)
h_c	Heat transfer coefficient from the inside of the tube to air (W/m^2K)
z	Depth (m)
A_s	Amplitude of soil surface variation $(T_{z, t})$ (K)
α_s	Soil thermal diffusivity (m^2/day)
α	Fraction of solar radiation absorbed by glass plate
t	Time elapsed from the beginning of the year (days)
t_o	Phase constant (days)
k_t	Thermal conductivity of tube $(W/m.\ K)$

Symbol	Meaning
k_i	Thermal conductivity of insulator $(W/m.\ K)$
K_p	Thermal conductivity of absorber plate $(W/m.\ K)$
K_{pcm}	Thermal conductivity of PCM layer $(W/m.\ K)$
h_1	Enthalpy of drying air at point 1 $(J/kg\ K)$
h_2	Enthalpy of drying air at point 2 $(J/kg\ K)$
V	Volt
AC	Alternating current
μ_w	water chemical potential
μ_{ow}	Standard state chemical potential
R	Gas constant
T	Absolute temperature
$U_{t,\,EAHX}$	Overall heat transfers coefficient of EAHX (W/m^2K)
$Q_{h,\,EAHX}$	Heat transfer rate of the EAHX (W)
ΔT_{lm}	Logarithmic average temperature difference
A_{EAHX}	Area of the heat exchanger (m^2)
K_a	Thermal conductivity of air $(W/m.\ K)$
δ_p	Thickness of absorber plate (m)
δ_{pcm}	Thickness (m)
δ_i	Thickness of the insulation (m)
Δx	Thickness of the wall of the geothermal heat exchanger pipe
$h_{c,\,p-a}$	Convective heat transfers coefficient between absorber plate and air (W/m^2K)
$h_{c,\,g-a}$	Convective heat transfers coefficient between glass and air (W/m^2K)
$h_{c,\,i-w}$	Convective heat transfers coefficient from insulator to wind (W/m^2K)
$h_{c,\,g-w}$	Convective heat transfers coefficient between glass and wind (W/m^2K)
$h_{r,\,g-s}$	Radiation heat transfers coefficient between glass and sky (W/m^2K)
$h_{r,\,g-p}$	Radiation heat transfers coefficient between glass and absorber plate (W/m^2K)
$h_{r,\,p-pcm}$	Radiation heat transfers coefficient between absorber plate and PCM (W/m^2K)
T_{out}	Desired outlet air temperature from EAHX (K)
T_{am}	Ambient air temperature (K)
T_{pcm}	Temperature of PCM layer (K)
$\dot{m}_a$	Mass flow rate of air (kg/s)
C_p	Specific heat of air $(J/kg.\ K)$
G	Direct Normal Irradiance (Wm^{-2})
$dA_{PA-FPSC}$	Small cross-sectional area of solar collector (m^2)
τ	Fraction of solar radiation absorbed by metal absorber plate

Chapter 1
Insights of Drying

Food is absolutely paramount for the survival of every human being [1]. Therefore, 'Right to food' is universally and unequivocally recognized as a fundamental human requirement through article 25 of 'Universal Declaration of Human Rights' adopted in 1948 [2]. However, regardless of the unprecedented technological achievements made in the twenty-first century, 'hunger' has remained as one of the most predominant challenges. Food waste plays a pivotal role in the world hunger crisis as roughly 1.3 billion tons (30% of produced food worldwide) of food are being wasted in the food chain every year [3–5]. As a result, one in three people worldwide remain malnourished, and 815 million people are prone to hunger on a daily basis [6]. Food waste can be defined as the loss of food in subsequent stages of the food supply chain intended for human consumption [7]. Therefore, ensuring the availability of food through the incorporation of food preservation technologies, is one of the prime demands of the modern era [8, 9]. While the availability of food and its wastage differs from nation to nation, all the developed countries follow an approximately similar waste trend. In contrast, a different trend can be observed in the developing countries, which is discussed later in this chapter. There are 195 UN recognized sovereign states in the world, and all of them have distinctive characteristics in terms of economic status, agricultural practices, food waste scenario, and hunger dimensions, etc. In order to continue the discussion on food waste and its proper management, countries can be categorized based on their GDP (PPP) per capita. Countries of the world can be classified in four groups based on the GDP (PPP) per capita: low income ($955 or less), Lower Middle Income ($966–$3895), Higher Middle Income ($3896–$12,055) and Higher Income ($12,056 to more). The authors recognize the countries of low income and lower-middle-income groups as developing counties. Considering the scope of this book, readers can expect to find detailed discussion on the following topics:

1. Current Energy status (Renewable and Non-Renewable)
2. Drying Techniques currently practiced

© Springer Nature Switzerland AG 2020
M. Hasan Masud et al., *Sustainable Food Drying Techniques in Developing Countries: Prospects and Challenges*,
https://doi.org/10.1007/978-3-030-42476-3_1

3. Proposed Drying Techniques for the concerned developing Countries
4. Constraints in implementing the proposed drying Technique

Chapter 1 of this book discusses the food waste problem, development and economic (e.g. GDP per capita) performance indicators of the developing countries, provides an overview of food supply chain and describes principles of food drying.

1.1   Developing Countries and their Economic Status

GDP (PPP) per capita is one of the most important indicators by which the economic condition of a country can be assessed. A box and whisker plot of GDP (PPP) per capita for four economic country groups of the world is presented in Figs. 1.1a

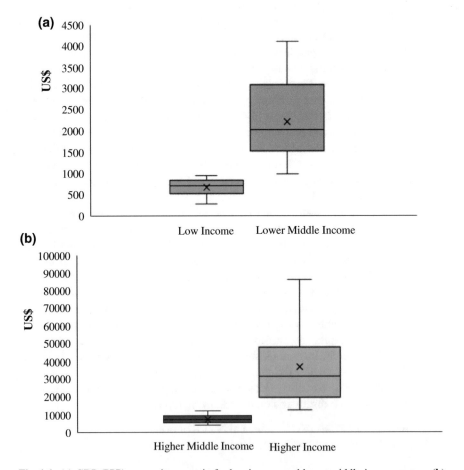

Fig. 1.1 (**a**) GDP (PPP) per capita scenario for low income and lower middle-income groups (**b**) GDP (PPP) per capita scenario for higher income and higher middle-income groups

and b. The plots were prepared based on the GDP (PPP) per capita from 195 countries. In low income and lower-middle-income categories, there were 27 and 55 countries, respectively, whereas, in the case of higher middle income and top income region, there were 62 and 51 countries, respectively. The analysis considers the individual GDP (PPP) of the 195 countries, along with the impact of international economic regions and trade pacts.

Considerable variation in the GDP (PPP) per capita can be found amongst the developing countries (low income and lower middle income) when compared with the developed countries' (Higher Middle Income and High Income) GDP (PPP). From Figs. 1.1a and b it can be observed that the average GDP (PPP) per capita for the low income and middle-income developing countries are US$ 674 and US$ 2231.77, respectively, whereas the average GDP (PPP) per capita for the higher middle income and higher income developed countries are US$ 7435.99 and US$ 37529.23, respectively.

From the data of 195 countries, it was found that the highest GDP (PPP) per capita in the higher middle-income group is in China (US$ 86355.40), and the lowest GDP (PPP) per capita in the low-income group is in Burundi (US$ 275.42). The median value of low income, lower-middle income, higher middle income, and higher income is US$ 712.45, 2021.88, 7233.99, and 31651.34, respectively. The spread of GDP (PPP) for countries in the various economic classifications are apparent from the box and whisker chart. Majority of the countries in the low-income category are below the median, whereas in the case of the lower-middle income group, a significant number of countries are over the median line. This is indicative of the high development potential of the middle-income countries in contrast to the low-income category. In the higher middle-income spectrum, the countries are almost evenly distributed across the median line. Similar to the middle-income category, more countries are placed above the median line in high-income criteria, which displays its positive growth and sustainability tread.

A synopsis of agricultural features, hunger dimensions, and food supply scenario for a set of selected developing (low and lower middle income) countries are represented in Figs. 1.2, 1.3, 1.4, respectively (note that: statistics for some individual years are unavailable in the annual reports) [10, 11]. While Fig. 1.2 indicates that in developing countries, the harvested area increased from 38.2% to 40.6% in 2016; the employment and added value in agriculture, forestry, and fishing followed a decreasing trend. On the other hand, Fig. 1.3 displays a decreasing trend in the following indicators- dietary energy supply, the prevalence of undernourishment (%), and underweight children percentage (children under five). It is alarming that food export is decreasing, and food import is increasing in developing countries, which is also indicated in Fig. 1.4. The agricultural scenario of the developing countries is riddled with inequalities. Even though the majority of the human resources and maximum of the agricultural land is utilized in developing countries, due to the ineffective cultivation and post-harvest food processing systems, the scenario of food waste is worse in developing countries. Moreover, the exorbitant cost of food production in developing and least developed countries is only exacerbated by the

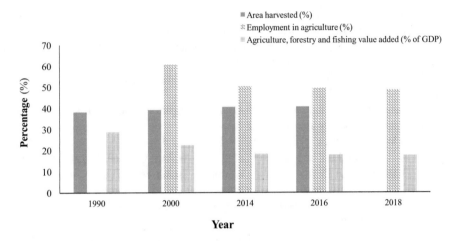

Fig. 1.2 Synopsis of an agricultural aspect of developing countries

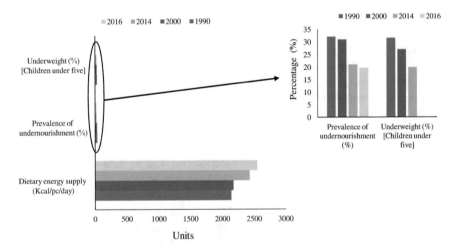

Fig. 1.3 Hunger dimensions of developing countries

excessive amount of food waste. This phenomenon eventually leads to more import of food than export when compared to the developed countries.

With the steady growth of the world population, food waste management has become a top priority of national interest in several countries. People in the developing nations spend roughly 50–80% of their income on food; thus, a continuous rise in the price of food can lead a nation to economic jeopardy [12]. Food production is very uneven throughout the world due to geographical differences and agricultural conditions. Due to the complexities involved in the mobility of food, it is very difficult and costly to transport the surplus fresh food into other places. Even with the best preservation technologies available, the quality of food degrades with time.

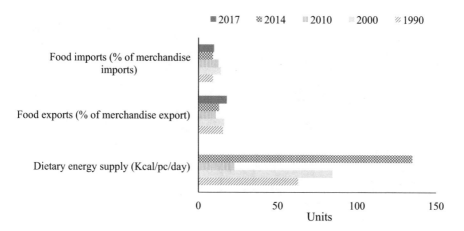

Fig. 1.4   Food supply scenario of developing countries

1.2   Food Waste Scenario

In order to understand the necessity of food preservation, particularly food drying, to minimize the amount of food waste, it is necessary to have a clear picture of the food waste scenario. The amount of wastes of Fresh fruits, vegetables, meat, bakery, and dairy products along the supply chain varies from developed to developing countries [13, 14]. Nearly 1.3 billion tons of perishable food is wasted along the food supply chain [15–17]. Food waste from production to the retailer stage is high in developing countries compared to developed countries because of insufficient use of technologies and inadequate infrastructures. On the contrary, food waste in consumer stage is much higher in developed countries than in developing nations. Factors that influence food waste include consumer psychology, food quality, along with some other features that contribute to health and lifestyle balance [18, 19]. Wastage of food in various phases of food supply chain occurs at a disproportionate rate. Different stages of food waste are represented in Fig. 1.5.

The proportions may vary due to the differences in operating conditions found in different regions of the globe. In general, maximum losses occur in the production phase in all the countries, which is about 25~38 percent of the overall food loss. Among the factors contributing to food waste, premature harvest, and adverse climatic situations have the most profound effect. Post-harvest loss amounting 10~25% of the total food loss occurs primarily because of humid weather, improper packagings, and inadequate food storage systems. Due to the utilization of advanced storage facilities, the share of post-harvest loss is lower in developed countries compared to their developing counterparts.

Due to the high seasonality of foods, along with poor preservation techniques and improper processing, 5~15% of food is being wasted. The other losses occur in the consuming stage accounting for 10~40% of the total food waste around the world. The severity and inefficiency of waste in the postharvest handling and

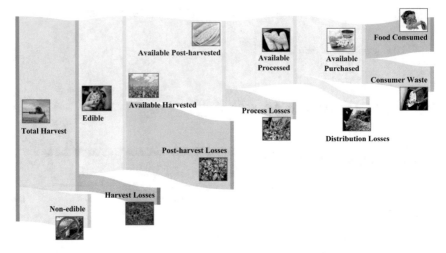

Fig. 1.5 Food losses scenario in the supply chain

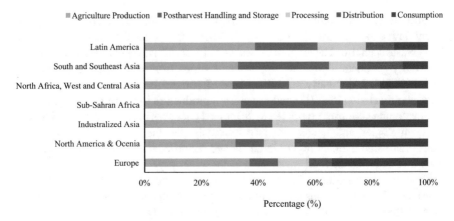

Fig. 1.6 Mosaic plot of food losses occurred in subsequent stages of the food supply chain across the globe (Adapted from Joardder et al. [10])

processing stages are more significant in developing countries compared to the developed countries.

The mosaic plot of the food supply chain in Fig. 1.6 explicitly shows that in developing regions like South and Southeast Asia, North Africa, West, and Central Asia food waste mostly occurs in the post-harvest handling & storage and processing stages. Essentially, upgrading these stages with advanced techniques will undoubtedly reduce the severity of food waste.

Food waste alone accounts for 45% of total municipal solid waste in Europe [20]. This scenario is direr in the developing and least developed nations where, on average, 55% of the total municipal solid waste is food [21–24]. Given the severity of food shortage and its impact on society, measurements for controlling and managing food waste has been practiced collectively since the 1900s. The first remarkable movement is being initiated as early as 1896 in New York, where people started separating food form their solid waste [25]. Major steps towards ensuring adequate food and its waste management were taken after the event of World War II. It started with the creation of the Food and Agricultural Organization (FAO) by the United Nations (UN) in 1945. A pioneering example of international collaboration for food security was the efforts to recognize 'right to food' under the universal declaration of human rights. Conscious food waste management was introduced in the 1974 World Food Conference in Rome by examining the food production and consumption process. These agendas were ratified in the 1996 World Summit with clear action plans. In the early twenty-first century, with an adoption of the millennium developing goal (MDG), the world, for the first time, saw a unified global approach towards tackling food waste.

Consumers in developed countries like Europe and North-America waste 95~115 kg/year, which is about 40% of their total food [26, 27]. In contrast, consumers of developing countries like sub-Saharan Africa and South/Southeast Asia waste only around 6~11 kg/year, which accounts for 6% of their total food. This is a glaring issue of the contrasting food loss behavior among the developing and developed nations. Consequently, the agriculture-related challenges faced in the stages from production to the retailing is exacerbating the food waste scenario in developing countries. In case of developing countries, storage problem is the foremost barrier to an efficient food supply chain management. Research has shown that effective and improved drying technologies abate the storage problems. Therefore, it is highly recommended to introduce drying as a strategy to ameliorate the appalling food loss problem in developing countries. Improved drying technologies have the potential to initiate groundbreaking improvement in post-harvest handling, storage, and processing stages of the food chain. This book has proposed several sustainable drying techniques that are deemed feasible for developing countries.

1.3 Food Drying

Drying is one of the oldest and the most efficient approaches to food processing [28]. Inadequate storage facilities and improper processing of perishable foods are responsible for almost one-third of the produced food waste throughout the world [8, 9]. The scenario is aggravated in the developing countries, where approximately 30–40% perishable food materials are wasted due to the lack of timely and proper processing [29, 30]. Drying can certainly be recommended as a simple and extensively used technique to overcome this alarming issue of food waste in the processing stage. Hot air or convective drying (CD) is the easiest and oldest method of food drying [31]. However, drying is one of the most energy-intensive processes [32–34]. Most of the drying in developing countries is carried out either under the sun or using hot air. In some developing as well as developed countries, different types of solar drying systems are also used [23, 35, 36]. However, in developed countries, numerous improved drying techniques are practiced, such as drum drying, spray drying, freeze-drying, and Infrared radiation drying, etc. [37, 38]. All of these methods raise concerns about either its high energy requirement or quality management [39, 40]. The quality of dried food is affected by the state of fresh food, method of preparation, and drying conditions [41, 42]. The issue of the energy requirement and quality of dried food has received considerable attention in recent years. Since the moisture content of most of the fresh fruits and vegetables ranges between 80–95%; therefore, they are classified as highly perishable commodities [30, 43, 44]. Large amount of fresh fruits and vegetables are being wasted primarily due to lack of proper preservation. The main reason behind their perishability is the high water content present in these fresh fruits and vegetables. Along with cold storage, drying is also practiced to preserve these perishable fresh fruits and vegetables across the globe. Considering the substantial amount of water present in perishable fresh fruits and vegetables, it is essential to have detailed knowledge about the water distribution in the food materials.

1.3.1 Water Distribution in Fresh Food

Water distribution inside the food tissue significantly affects the degree of structural changes. Depending on the properties, water in food materials may be categorized into three types, namely free water, loosely bound water, and strongly bound water [45, 46]. Based on the distribution of water inside the food tissue, water can also be classified as cell wall water, intercellular water, and intracellular water (See Fig. 1.7). An overview of different types of water present in food tissue is given below.

1.3.1.1 Free Water

Free water is generally a mixture of liquid and gas and is typically located in the intercellular spaces. Without subsequent shrinkage, free water eases the high-temperature decomposition reaction and internal transport of compounds during drying. After the removal of all the free water, deterioration of food quality, and high energy requirement for removal of bound water is experienced. Free water is also termed as bulk water and capillary water, depending on the nature of food material [47].

1.3.1.2 Bound Water

Bound water has no standard definition; however, it basically demonstrates different properties when compared to the "free water," and this is the type of water that is strictly attached to other compounds of food tissue. There are a few techniques to

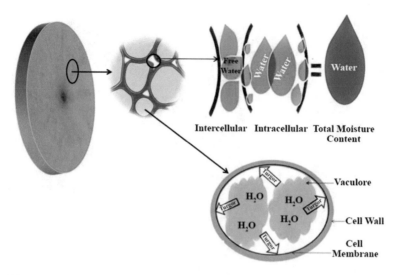

Fig. 1.7 Water distribution within plant-based cellular tissue

measure bound water that is comprehensively discussed in numerous literature [48]. As bound water is firmly bound, a high amount of heat is required during its separation from the non-water constituents [49]. Bound water can also be classified into two types; loosely bound water and strongly bound water. In another way, the bound water is classified as physically bound water and chemically bound water.

1.3.1.3 Spatial Water Distribution

Water can also be categorized depending on the spatial location of the existence of water in food tissue. In plant-based cellular tissue, water can exist in three different regions- intercellular spaces, intracellular spaces, and in the cell wall (See Fig. 1.7).

On wet basis, there is around 78–97% water available in intracellular spaces, and the remaining is the intercellular water and cell wall water. Although it is complicated to determine the share of water in cell walls and cells, some of the researchers have attempted to quantify the proportion of water [48, 50–53]. There are many distinguishing features that differentiate the intracellular, intercellular, and cell wall water, which are comprehensively represented in Fig. 1.8.

The requirement of energy and the optimization of different drying parameters are significantly affected by the quantity and quality of water in the dissimilar cellular environment [50].

Intercellular water
- Water of intercellular spaces
- Present in the form of crevices, voids and capillaries
- 5 to 15% of the total water (wet basis)
- Vaporize at the early stage of drying

Intracellular Water
- Water within cell and cell wall
- 78-97% of the total water (wet basis)
- Cell wall is ruptured when this type of water is removed
- Pore foramtion and cell collapse also occured when this type of water is vaporized

Cell wall water
- Present in the cell wall
- Also known as strongly bound water or extracellular water
- Needs maximum amount of energy to remove this type of water

Fig. 1.8 Characteristics of different types of water

1.3.2 Mass Transfer During Drying

Drying is basically a simultaneous heat and mass transfer process. Therefore, in order to understand the underlying mechanism of drying, it is obvious to have a considerable amount of knowledge on the water migration pathways and mechanisms in food materials. A significant volume of moisture is transferred during drying that results in cell collapse, pore formation, and many other changes. Transmembrane transport (across cell membranes), apoplastic transport (cell wall to free space), and symplastic transport (movement of water between two neighboring cells) are the three possible techniques by which water can be transported as proposed by Le Maguer and Yao [54].

The presence of three different types of water flux, namely cell-to-cell, intercellular, and wall-to-wall, is reported by Crapiste et al. [55]. In cell-to-cell, water fluxes through cell membranes, vacuoles, and cytoplasm. In the case of the wall to wall, capillary flux across cell walls is observed, whereas vapor movement through intercellular spaces is observed in case of the third type of water flux. Moreover, Joardder et al. reported that there are different pathways (intercellular, cells, and cell wall zones) by which the moisture can migrate to the surroundings in the time of drying [56]. Considering the above-mentioned facts, a water migration pathway is revealed in Fig. 1.9.

The water of different regions may migrate through various pathways, as represented in Fig. 1.9. For instance, if no cell collapse occurs in the beginning stage of the drying, then water present in the Intracellular spaces follows the migration

Fig. 1.9 Conceptual maps
of different pathways of
water migration from
plant tissue

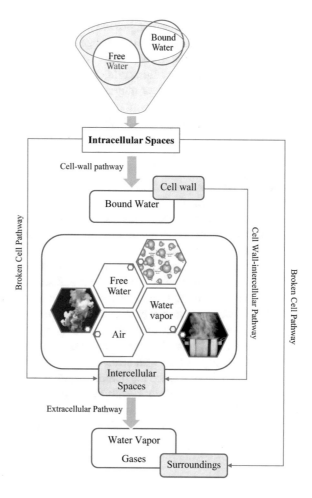

pathway-Cell wall to intercellular to extracellular to surroundings. However, if cell collapse occurs in the initial stage of drying, then the water migration follows the broken cell pathway, which is-Intercellular spaces to surroundings. The first pathway (Cell wall to intercellular to extracellular to surroundings) is followed as long as the cell walls remain intact (at a lower temperature). However, when the temperature is increased, the cell membrane and walls are ruptured, and consequently, the water migrates through the broken cell pathway. Compared to the migration of intercellular and intracellular water, as mentioned earlier, removal of cell wall water requires a higher amount of energy as it moves through the inter-microfibrillar regions that are around 10 nm in cross-section and several times lengthier [57].

Due to the water migration from the plant tissues, a wide range of structural modifications in the food samples during drying may occur. A step-by-step structural modification process due to the mass transfer is presented here [26, 27, 47, 58–60].

1. In the initial stage of drying, just the intercellular (free) water is evaporated to the surrounding and involves no shrinkage of the sample.
2. The second stage is known as the falling rate stage, where pore shrinkage occurs due to the migration of intracellular water from the cell. Moreover, in the early stage of intracellular water migration, the magnitude of volumetric shrinkage is actually similar to the amount of migrated water.
3. As a result of the migration of intercellular water and lower turgor pressure, cell shrinkage occurs. Cell reduction is expected to be proportional to the ratio of water density to particle density. Hence, due to water migration, a void portion is developed, which remains uncontracted and increases dried product porosity. Particle density is assumed to be within the range between water density and solid product density [61].
4. In most cases, initial shrinkage takes place without cell collapse. Although cell breakage occurs due to longer drying time, it doesn't cause severe cell collapse, particularly at the beginning. Migration of cell wall water is the main reason behind cell collapse [62]. The most thin-walled cell in the plant-based food materials is parenchyma cell. The cell that stores water is known as a vascular cell, which is located inside the cell. Vascular solutes are usually osmotically active, which means they push cell membranes against the cell walls.
5. Consequently, the turgor pressure increases and keeps the cells in a state of elastic stress maintaining its shape, firmness, and crispness of tissue [63]. With the loss of turgor pressure, fruit structure is lost, and lot of microstructural changes take place [64–67]. Once the natural turgidity is lost, it cannot be restored [68].
6. In the final stage of drying, both the collapse of the cells and shrinkage of the food solid matrix are hindered as they are adequately dried, and at the same time, their viscosity also becomes higher [69, 70]. Moreover, the intercellular spaces of the food materials enlarge if drying is continued for a prolonged duration. The physical interpretation of pore evolution during drying is represented in Fig. 1.10 [56].

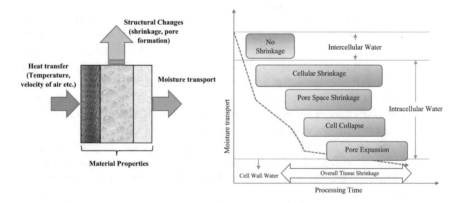

Fig. 1.10 Physical interpretation of pore evolution during the time of drying

Shape modification and deformation during drying occur because of the stress developed due to the simultaneous change in moisture and thermal gradients [71]. Moisture gradient stress plays the leading role in generating stress during drying, as after a certain period of drying, the thermal stress becomes insignificant [72, 73]. Therefore, the moisture gradient is the primary reason behind the plant tissue shrinkage.

Nevertheless, dissimilar categories of internal moisture transfer mechanisms may occur during drying, which entirely depends on the sample properties and process condition of drying. Different methods of moisture migration reported in the literature are summarized in Table 1.1.

It can be seen in Table 1.1 that the drying process involves several mass transfer mechanisms. However, during drying, only one of those mechanisms may dominate at a time given time [20]. In some instances, moisture migration within the food tissue might happen by more than one of the above-mentioned mechanisms of mass transfer.

1.3.3   Heat Transfer During Drying

Drying is a simultaneous heat and mass transfer process with a high degree of complexity. Several types of physical or chemical transformations may occur during drying that subsequently might change the rate of heat and moisture transfer. Moreover, significant changes may also occur in the product quality that includes puffing, glass transitions, shrinkage, and crystallization.

Drying is fundamentally accomplished by the effective vaporization of the moisture present in the food issue. When a substantial amount of heat is applied, moisture migration starts following different pathways. However, depending on the type of drying process, different modes of heat transfer may come into consideration. In direct dryers, convection heat transfer is occurred, in contact or indirect dryer's conduction heat transfer is occurred, and in dielectric (microwave and radio frequency) dryers, volumetric heating is accomplished. At the industrial level, approximately 85 percent of dryers are convective types. In microwave and radio frequency drying, the microwave directly passes through the food tissue and forces the water molecule to vibrate faster (See Fig. 1.11). All the other modes of drying provide heat at the boundaries of the food tissue in order to diffuse heat into the solid matrix of the food material, where heat is primarily transferred by conduction process [84]. When heat is conducted through the food tissue, moisture is migrated via the principle of diffusion from the food tissue to its surroundings. The stages of drying, along with its heat transfer mechanism, is shown in Fig. 1.11.

In the first stage of drying, as free moisture is available on the surface, the vaporization of those water particles occurs very quickly, and a persistent moisture removal rate is observed [85]. Different types of external attributes assist in removing moisture at this stage of drying that includes external mass and heat transfer coefficients, exposed area to the dry air, and the temperature gradient between dry

Table 1.1 Internal moisture transfer mechanisms during drying of solid foods

Moisture removal mechanism during drying		Reference
Diffusion		[74–76]
Capillary		[77]
Evaporation-condensation		[78]
Transfer of liquid water	Removal of water vapor	[79]
Capillary	Differences in partial pressure (diffusion)	
Liquid diffusion		
Surface diffusion	Differences in total pressure (hydraulic flow)	
Capillary Diffusion Surface diffusion in liquid layers absorbed at solid interfaces Water vapor diffusion in air-filled pores Gravity Vaporization-condensation sequence.		[80]
Hydraulic flow Capillary flow Evaporation-condensation Vapor diffusion		[81]
Molecular diffusion Capillary flow Knudsen flow Hydrodynamic flow Surface diffusion		[82]
Transfer of liquid water Capillary (saturated)	Transfer of water vapor Diffusion (in pores):	[82]
Molecular diffusion (within solid) Surface diffusion (absorbed water) Liquid diffusion (in pores) Hydraulic flow (in pores)	- Knudsen, - Ordinary, -Stephan diffusion.	
	Hydraulic flow (in pores) Evaporation-condensation	
Transfer of liquid water	Transfer of water vapor	[83]
Capillary Diffusion Surface diffusion Hydraulic flow	Mutual diffusion Knudsen flow Diffusion Slip flow Hydrodynamic (bulk) flow Stephan diffusion Poiseuille flow Evaporation-condensation	

air and wet surface [86]. The initial stage of drying is also known as the surface-based drying.

After the initial stage, the constant moisture removal rate is obstructed, and in the following stage, the diffusion of moisture from the inside of the food material surface is required. This stage of drying is known as the first falling rate, and the amount of moisture content in this stage is known as critical moisture content (MC_c). The shape of the food material sample, properties of the material, and

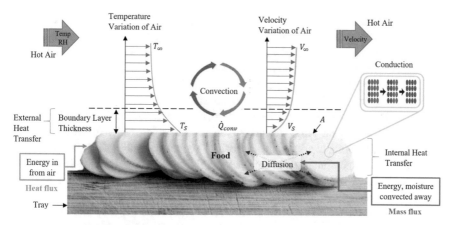

Fig. 1.11 Conceptual map of heat transfer during drying

thickness are the key factors influencing the drying rate. In the later stage of drying, owing to the lack of moisture, heat damage might occur in the surface region. Due to heat flux reduction and supplementary internal resistance for moisture transfer, the rate of drying is reduced in that stage.

The last stage of drying is known as the second falling rate of drying. Owing to the concentration gradient present between the surface and the core region, diffusion occurs inside the food tissue that subsequently results in the migration of moisture from the center to the surface. In this phase of drying, the food material sample may experience longer drying time, due to the lower moisture gradient. The first 90% of the moisture removal and the last 10% of the moisture removal takes an almost equal amount of time. Most of the food materials experience this second falling rate of drying. The drying rate can be accelerated by providing more heat energy. However, a hard crust may form due to a higher temperature during drying, which may subsequently damage the surface of the dried food material sample [87]. As the conductivity of dried/hard food tissues is inferior; therefore, an excess supply of heat may damage the surface of the food materials by overheating. Heat and mass transfer rates may be hindered by this damaged surface [88].

1.3.4   Relation Between Process Parameters of Drying and Food Quality

Dried food qualities are controlled by several drying parameters and material properties. Moreover, there are also some imperative factors, including porosity level, puffing, glass transitions, shrinkage, case hardening, and crystallization, that significantly affect the desired quality of the dried food. By controlling the process parameters, the level of porosity and all the other factors can be controlled, which may subsequently result in high quality dried food. The bridge between the quality of dried food and drying factors is porosity [56]. The hypothesized relationship between drying parameters and dried food quality is presented in Fig. 1.12.

Porosity impacts water migration to a considerable degree, which is the key action to accomplish the drying process. Among the factors on which the porosity level depends are sample shape, temperature, coating treatment, and mobility of the solid. Conversely, the drying rate and pressure of the drying system have an inversely proportional relationship with the porosity. Therefore, establishing a successful relationship between the process parameters of drying and food quality is one of the prime objectives of drying research. The following benefit may be obtained by establishing such a relationship (see Fig. 1.13).

Understanding the fundamental relationship between process parameters and product quality and determining process parameters based on this knowledge will significantly advance the food processing system that can minimize the food waste. Further technological advancement may provide ways to improve the quality of drying in the developing countries.

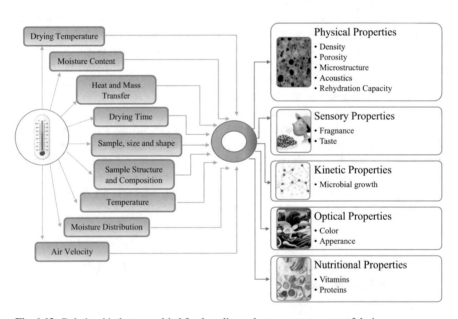

Fig. 1.12 Relationship between dried food quality and process parameters of drying

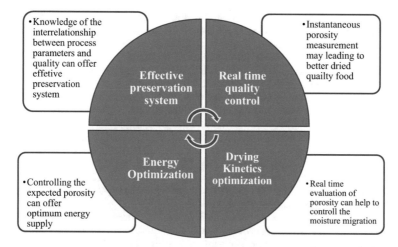

Fig. 1.13 Outcomes of establishing the interrelationship between process parameters and dried food quality

References

1. C. Kumar, M.A. Karim, Microwave-convective drying of food materials: A critical review. Crit. Rev. Food Sci. Nutr. **59**(3), 379–394 (2019)
2. U. Nations, "Universal Declaration of Human Rights," (2015)
3. W. H. Organization, International decade for action water for life, 2005–2015. Wkly. Epidemiol. Rec. Relev. épidémiologique Hebd. **80**(22), 195–200 (2005)
4. M.I.H. Khan, S.A. Nagy, M.A. Karim, Transport of cellular water during drying: An understanding of cell rupturing mechanism in apple tissue. Food Res. Int. **105**, 772–781 (2018)
5. M.I.H. Khan, T. Farrell, S.A. Nagy, M.A. Karim, Fundamental understanding of cellular water transport process in bio-food material during drying. Sci. Rep. **8**(1), 15191 (2018)
6. FAO, *The Right to Food,* United Nations (2015)
7. FAO, *Food Loss and Food Waste,* United Nations (2016)
8. C. Kumar, M.U.H. Joardder, T.W. Farrell, G.J. Millar, M.A. Karim, Mathematical model for intermittent microwave convective drying of food materials. Dry. Technol. **34**(8), 962–973 (2016)
9. C. Kumar, M.U.H. Joardder, T.W. Farrell, M.A. Karim, Investigation of intermittent microwave convective drying (IMCD) of food materials by a coupled 3D electromagnetics and multiphase model. Dry. Technol. **36**(6), 736–750 (2018)
10. M. M. H. Joardder M.U.H, Foods and developing countries, in *Food Preservation in Developing Countries: Challenges and Solutions*, Springer International Publishing, Cham (2019)
11. FAO, "Statistical Yearbook," (2018)
12. UNDESA, "Water and Food Security- International Decade for Action 'Water for Life' 2005–2015," (2014)
13. M. Balaji, K. Arshinder, Modeling the causes of food wastage in Indian perishable food supply chain. Resour. Conserv. Recycl. **114**, 153–167 (2016)

14. M.U. Joardder, S. Mandal, M.H. Masud, Proposal of a solar storage system for plant-based food materials in Bangladesh. Int. J Amb. Eng. 1–17 (2018)
15. M.U.H. Joardder, M.H. Masud, Causes of food waste, in *Food Preservation in Developing Countries: Challenges and Solutions*, Springer, 27–55 (2019)
16. M.I.H. Khan, M.A. Karim, Cellular water distribution, transport, and its investigation methods for plant-based food material. Food Res. Int. **99**, 1–14 (2017)
17. M.I.H. Khan, C. Kumar, M.U.H. Joardder, M.A. Karim, Determination of appropriate effective diffusivity for different food materials. Dry. Technol. **35**(3), 335–346 (2017)
18. A.V. Cardello, H.G. Schutz, L.L. Lesher, Consumer perceptions of foods processed by innovative and emerging technologies: A conjoint analytic study. Innov. Food Sci. Emerg. Technol. **8**(1), 73–83 (2007)
19. C. Bonaui et al., Food drying and dewatering. Dry. Technol. **14**(9), 2135–2170 (1996)
20. IPCC, *National Greenhouse Gas Inventories,* Geneva, Switzerland (2006)
21. A.M. Troschinetz, J.R. Mihelcic, Sustainable recycling of municipal solid waste in developing countries. Waste Manag. **29**(2), 915–923 (2009)
22. M. Mourshed, M.H. Masud, F. Rashid, M.U.H. Joardder, Towards the effective plastic waste management in Bangladesh: A review. Environ. Sci. Pollut. Res. **24**(35), 27021–27046 (2017)
23. M.A. Karim, M.N.A. Hawlader, Performance investigation of flat plate, v-corrugated and finned air collectors. Energy **31**(4), 452–470 (2006)
24. M.H. Masud et al., Towards the effective E-waste management in Bangladesh: A review. Environ. Sci. Pollut. Res. **26**(2) (2019)
25. K. Coopersmith, *A Brief History of Food Waste,* Food Mentum Magazine (2016)
26. M.U.H. Joardder, C. Kumar, M.A. Karim, Prediction of porosity of food materials during drying: Current challenges and directions. Crit. Rev. Food Sci. Nutr. **58**(17), 2896–2907 (2018)
27. M.U.H. Joardder, M.A. Karim, Development of a porosity prediction model based on shrinkage velocity and glass transition temperature. Dry. Technol., 1–17 (2019)
28. T. Koyuncu, İ. Tosun, Y. Pınar, Drying characteristics and heat energy requirement of cornelian cherry fruits (Cornus mas L.). J. Food Eng. **78**(2), 735–739 (2007)
29. M.A. Karim, M.N.A. Hawlader, Drying characteristics of banana: Theoretical modelling and experimental validation. J. Food Eng. **70**(1), 35–45 (2005)
30. M.A. Karim, M.N.A. Hawlader, Mathematical modelling and experimental investigation of tropical fruits drying. Int. J. Heat Mass Transf. **48**(23), 4914–4925 (2005)
31. I. Alibas, Microwave, air and combined microwave–air-drying parameters of pumpkin slices. LWT-food Sci. Technol. **40**(8), 1445–1451 (2007)
32. C. Kumar, M.U.H. Joardder, A. Karim, G.J. Millar, Z. Amin, Temperature redistribution modelling during intermittent microwave convective heating. Procedia Eng. **90**, 544–549 (2014)
33. M.H. Masud, M.U.H. Joardder, M.A. Karim, Effect of hysteresis phenomena of cellular plant-based food materials on convection drying kinetics. Dry. Technol. **37**(10) (2019)
34. C. Kumar, M.U.H. Joardder, T.W. Farrell, M.A. Karim, Multiphase porous media model for intermittent microwave convective drying (IMCD) of food. Int. J. Therm. Sci. **104**, 304–314 (2016)
35. M.A. Karim, M. Hawlader, Development of solar air collectors for drying applications. Energy Convers. Manag. **45**(3), 329–344 (2004)
36. M.H. Masud, R. Ahamed, M. Mourshed, M.Y. Hossan, M.A. Hossain, Development and performance test of a low-cost hybrid solar air heater. Int. J. Ambient Energy **40**(1) (2019)
37. C. Kumar, G.J. Millar, M.A. Karim, Effective diffusivity and evaporative cooling in convective drying of food material. Dry. Technol. **33**(2), 227–237 (2015)
38. C. Kumar, M.U.H. Joardder, T.W. Farrell, G.J. Millar, A. Karim, A porous media transport model for apple drying. Biosyst. Eng. **176**, 12–25 (2018)
39. N.D. Pham, W. Martens, M.A. Karim, M.U.H. Joardder, Nutritional quality of heat-sensitive food materials in intermittent microwave convective drying. Food Nutr. Res. **62** (2018)
40. N. Duc Pham et al., Quality of plant-based food materials and its prediction during intermittent drying. Crit. Rev. Food Sci. Nutr. **59**(8), 1197–1211 (2019)

41. V. Puranik, P. Srivastava, V. Mishra, D.C. Saxena, Effect of different drying techniques on the quality of garlic: A comparative study. Ame J Fd Tech **10** (2012)
42. M.U.H. Joardder, R.J. Brown, C. Kumar, M.A. Karim, Effect of cell wall properties on porosity and shrinkage of dried apple. Int. J. Food Prop. **18**(10), 2327–2337 (2015)
43. M.U.H. Joardder, M.A. Karim, C. Kumar, Better understanding of food material on the basis of water distribution using thermogravimetric analysis, in *International Conference on Mechanical, Industrial and Materials Engineering (ICMIME2013). Rajshahi, Bangladesh*, (2013)
44. V. Orsat, W. Yang, V. Changrue, G.S.V. Raghavan, Microwave-assisted drying of biomaterials. Food Bioprod. Process. **85**(3), 255–263 (2007)
45. Z. Welsh, C. Kumar, A. Karim, Preliminary investigation of the flow distribution in an innovative intermittent convective microwave dryer (IMCD). Energy Procedia **110**, 465–470 (2017)
46. Z. Welsh, M.J. Simpson, M.I.H. Khan, M.A. Karim, Multiscale Modeling for food drying: State of the art. Compr. Rev. Food Sci. Food Saf. **17**(5), 1293–1308 (2018)
47. M.U.H. Joardder, M. Mourshed, M.H. Masud, Water in foods BT, in *State of Bound Water: Measurement and Significance in Food Processing*, ed. by M. U. H. Joardder, M. Mourshed, M. H. Masud, Springer International Publishing, Cham, 7–27 (2019)
48. M.U.H. Joardder, M. Mourshed, M.H. Masud, Bound water measurement techniques BT, in *State of Bound Water: Measurement and Significance in Food Processing*, ed. by M. U. H. Joardder, M. Mourshed, M. H. Masud, Springer International Publishing, Cham, 47–82 (2019)
49. D.R. Briggs, Water relationships in colloids. II. J. Phys. Chem. **36**(1), 367–386 (1932)
50. M.I.H. Khan, R.M. Wellard, S.A. Nagy, M.U.H. Joardder, M.A. Karim, Investigation of bound and free water in plant-based food material using NMR T2 relaxometry. Innov. Food Sci. Emerg. Technol. **38**, 252–261 (2016)
51. Q.T. Pham, Calculation of bound water in frozen food. J. Food Sci. **52**(1), 210–212 (1987)
52. G. Velazquez, A. Herrera-Gómez, M.O. Martín-Polo, Identification of bound water through infrared spectroscopy in methylcellulose. J. Food Eng. **59**(1), 79–84 (2003)
53. M.U.H. Joardder, M. Mourshed, M.H. Masud, Significance of bound water measurement BT, in *State of Bound Water: Measurement and Significance in Food Processing*, ed. by M. U. H. Joardder, M. Mourshed, M. H. Masud, Springer International Publishing, Cham, 119–135 (2019)
54. M. Le Maguer, Z.M. Yao, Mass transfer during osmotic dehydration at the cellular level. Food Preserv. by Moisture Control Fundam. Appl., 325–350 (1995)
55. G.H. Crapiste, S. Whitaker, E. Rotstein, Drying of cellular material—I. A mass transfer theory. Chem. Eng. Sci. **43**(11), 2919–2928 (1988)
56. M. U. H. Joardder, A. Karim, C. Kumar, R. J. Brown, Porosity: Establishing the relationship between drying parameters and dried food quality, Springer International Publishing (2015)
57. P. P. Lewicki, A. Lenart, Osmotic dehydration of fruits and vegetables, In Handbook of Food Drying (1995)
58. M.U.H. Joardder, C. Kumar, R.J. Brown, M.A. Karim, A micro-level investigation of the solid displacement method for porosity determination of dried food. J. Food Eng. **166**, 156–164 (2015)
59. M. Mahiuddin, M.I.H. Khan, N.D. Pham, M.A. Karim, Development of fractional viscoelastic model for characterizing viscoelastic properties of food material during drying. Food Biosci. **23**, 45–53 (2018)
60. M. Mahiuddin, M.I.H. Khan, C. Kumar, M.M. Rahman, M.A. Karim, Shrinkage of food materials during drying: Current status and challenges. Compr. Rev. Food Sci. Food Saf. **17**(5), 1113–1126 (2018)
61. I.N. Ramos, T.R.S. Brandão, C.L.M. Silva, Structural changes during air drying of fruits and vegetables. Food Sci. Technol. Int. **9**(3), 201–206 (2003)
62. M.I.H. Khan, R.M. Wellard, S.A. Nagy, M.U.H. Joardder, M.A. Karim, Experimental investigation of bound and free water transport process during drying of hygroscopic food material. Int. J. Therm. Sci. **117**, 266–273 (2017)

63. R. Ilker, A.S. Szczesniak, Structural and chemical bases for texture of plant foodstuffs. J. Texture Stud. **21**(1), 1–36 (1990)
64. M.M. Rahman, M.U.H. Joardder, M.I.H. Khan, N.D. Pham, M.A. Karim, Multi-scale model of food drying: Current status and challenges. Crit. Rev. Food Sci. Nutr. **58**(5), 858–876 (2018)
65. M.M. Rahman, C. Kumar, M.U.H. Joardder, M.A. Karim, A micro-level transport model for plant-based food materials during drying. Chem. Eng. Sci. **187**, 1–15 (2018)
66. M.M. Rahman, Y.T. Gu, M.A. Karim, Development of realistic food microstructure considering the structural heterogeneity of cells and intercellular space. Food Struct. **15**, 9–16 (2018)
67. M.M. Rahman, M.U.H. Joardder, A. Karim, Science direct non-destructive investigation of cellular level moisture distribution and morphological changes during drying of a plant-based food material. Biosyst. Eng. **169**, 126–138 (2018)
68. R.M. Reeve, Relationships of histological structure to texture of fresh and processed fruits and vegetables. J. Texture Stud. **1**(3), 247–284 (1970)
69. J.M. Del Valle, T.R.M. Cuadros, J.M. Aguilera, Glass transitions and shrinkage during drying and storage of osmosed apple pieces. Food Res. Int. **31**(3), 191–204 (1998)
70. V.T. Karathanos, S.A. Anglea, M. Karel, Structural collapse of plant materials during freeze-drying. J. Therm. Anal. Calorim. **47**(5), 1451–1461 (1996)
71. L. Mayor, A.M. Sereno, Modelling shrinkage during convective drying of food materials: A review. J. Food Eng. **61**(3), 373–386 (2004)
72. P.P. Lewicki, G. Pawlak, Effect of drying on microstructure of plant tissue. Dry. Technol. **21**(4), 657–683 (2003)
73. K.S. Jayaraman, D.K. Das Gupta, N.B. Rao, Effect of pretreatment with salt and sucrose on the quality and stability of dehydrated cauliflower. Int. J. Food Sci. Technol. **25**(1), 47–60 (1990)
74. M.U.H. Joardder, A. Karim, C. Kumar, R.J. Brown, Determination of effective moisture diffusivity of banana using thermogravimetric analysis. Procedia Eng. **90**, 538–543 (2014)
75. S. Ghnimi, S. Umer, A. Karim, A. Kamal-Eldin, Date fruit (Phoenix dactylifera L.): An underutilized food seeking industrial valorization. NFS J. **6**, 1–10 (2017)
76. C.A. Perussello, C. Kumar, F. de Castilhos, M.A. Karim, Heat and mass transfer modeling of the osmo-convective drying of yacon roots (Smallanthus sonchifolius). Appl. Therm. Eng. **63**(1), 23–32 (2014)
77. N.H. Ceaglske, O.A. Hougen, Drying granular solids. Ind. Eng. Chem. **29**(7), 805–813 (1937)
78. P.S.H. Henry, The diffusion of moisture and heat through textiles. Discuss. Faraday Soc. **3**, 243–257 (1948)
79. P. Gorling, Physical phenomena during the drying of foodstuffs. Soc Chem Ind, 42–53 (1958)
80. van Arsdel, W. B., Copley, M. J., Morgan, A. I. (Eds.). Food Dehydration: Principles. AVI Publishing Company, Incorporated (1963)
81. R.B. Keey, P. Mandeno, T.K. Tuoc, Dissolution of freely suspended solid particles in an agitated non-newtonian fluid-rheology and mass transfer. Proc. Chemaca **70**, 53 (1970)
82. S. Bruin, K.C.A.M. Luyben, Drying of food materials: A review of recent developments, in *1st. Int. Symp. Drying, Montreal, 1978*, (1978)
83. K.M. Waananen, J.B. Litchfield, M.R. Okos, Classification of drying models for porous solids. Dry. Technol. **11**(1), 1–40 (1993)
84. A.S. Mujumdar, S. Devahastin, Fundamental principles of drying. Exergex, Bross. Canada **1**(1), 1–22 (2000)
85. A.S. Mujumdar, *Handbook of Industrial Drying,* CRC press (2006)
86. J. Srikiatden, J.S. Roberts, Moisture transfer in solid food materials: A review of mechanisms, models, and measurements. Int. J. Food Prop. **10**(4), 739–777 (2007)
87. C. Kumar, M.A. Karim, M.U.H. Joardder, Intermittent drying of food products: A critical review. J. Food Eng. **121**, 48–57 (2014)
88. G.E. Botha, J.C. Oliveira, L. Ahrné, Microwave assisted air drying of osmotically treated pineapple with variable power programmes. J. Food Eng. **108**(2), 304–311 (2012)

Chapter 2
Conditions for Selecting Drying Techniques in Developing Countries

Drying techniques are extensively practiced all over the world in order to alleviate the staggering volume of food waste in the processing stage [1, 2]. Lack of adequate post-harvest processing facilities exacerbates the food loss crisis, which is more widespread in developing countries [3, 4]. Effective moisture removal can elongate perishable foods' shelf life and ameliorate this abominable issue. Among the various approaches that are employed to remove moisture and preserve the food materials, drying is the most sustainable one [5]. Different types of drying technologies use distinctive thermodynamic principles, which should be carefully considered before proposing drying technologies to a specific developing country. The expected and unexpected quality attributes associated with the food drying process is shown in Fig. 2.1. The perceptual map implies that when process parameters such as minerals, protein, vitamin, taste, and fragrance are high, it produces high quality dried food; therefore, they are the expected food drying attributes.

Similarly, high quality dried food is produced when the microbial growth and color change is minimal. On the other end of the spectrum, when cost, temperature, energy, and drying time is high, dried food quality becomes undesirable. Hence, these attributes are unexpected. Subsequently, low rehydration capacity and low porosity are associated with poor quality dehydrated food. Key factors affecting the performance of drying techniques are represented in Fig. 2.2. However, while proposing drying techniques for developing countries, it may not be possible to consider all the factors. Therefore, the most significant attributes influencing the selection of drying techniques for developing countries will be critically discussed in this chapter.

During the drying process, the spatial water distribution primarily observed as bound or free plays a significant role. The removal of bound water requires a higher level of energy. However, excessive supply of heat energy over a prolonged period may render the food material unpalatable [6, 7]. This execrable situation occurs only when all the free water is migrated from the food tissue. Therefore, improved drying techniques focus on not only moisture removal but also on the retention of

© Springer Nature Switzerland AG 2020
M. Hasan Masud et al., *Sustainable Food Drying Techniques
in Developing Countries: Prospects and Challenges*,
https://doi.org/10.1007/978-3-030-42476-3_2

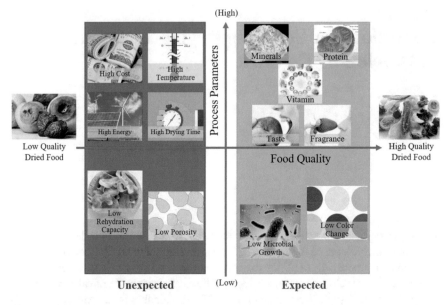

Fig. 2.1   Perceptual map of the expected and unexpected attributes of food drying

Fig. 2.2   Significant
attributes influencing the
selection of drying
techniques for developing
countries

the quality attributes at a satisfactory level. The losses of quality attributes include
the aroma loss, nutritive value loss, discoloring, losses of physical properties and
textural changes, etc. [8]. Processing time, along with the drying conditions, considerably affect the quality of dried foodstuffs [9]. For instance, drying at low temperatures may enhance the quality of the product, but the process will significantly
extend the drying time. Conversely, drying with high temperature reduces the duration of drying; but produces inferior quality dried food. Therefore, to propose any
drying technique, it is essential to optimize the processing time, quality attributes,
cost, safety, and energy requirement [10]. In the following section, a comprehensive
discussion on the different parameters for selecting drying techniques is presented
in detail.

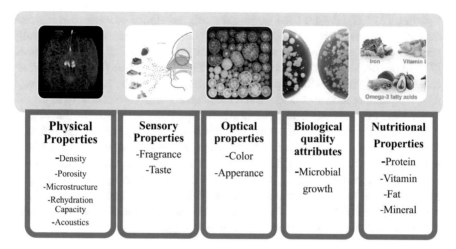

Fig. 2.3 Classification of food quality attributes

2.1 Quality Aspect of Drying

In general, quality is associated with the degree to which a product satisfies the requirements of the user [11]. In the context of food quality, it includes physical properties, sensory properties, optical properties, kinetic properties, and nutritional properties [12]. The most commonly used quality attributes are depicted in Fig. 2.3. Rahman and McCarthy (1999) categorized the qualities in four dominant sections, namely (1) kinetic properties, (2) sensory properties, (3) health properties, and (4) physical and physicochemical properties [13].

Food drying primarily affects the physical, nutritional, and microbial quality and stability. Depending on the moisture content and the mode of heat and mass transfer, the qualities may alter substantially. Overview of different imperative quality attributes of food materials is discussed in the following sections.

2.1.1 Physical Quality Attributes

2.1.1.1 Structure

Simultaneous heat and mass transfer phenomena substantially affect the microstructure of food material during drying [14–16]. Numerous researchers have tried to identify the influence of heat and mass transfer on the microstructure of non-porous and porous materials [17]. Structurally, food materials are classified as porous materials; therefore, they have a complex water transport mechanism in contrast to the non-porous materials [18–20]. Water migration inside the food tissue following the various pathways encounters a significant amount of resistance due to its porous structure [21, 22]. Before food drying, the cell membrane of each food material normally remains intact [17]. When the drying begins, the microstructure gradually starts to deteriorate, which consequently influences the rate of moisture removal [23]. In the beginning, drying does not cause significant damage to the microstructure; therefore, the structure is almost intact [24]. Conversely, as the drying progress, temperature increases, and the cell membrane is broken drastically [25].

As more than one mechanism of moisture migration may be involved in drying, it is challenging to predict the structure modification during its process [26, 27]. Consequently, it is indispensable that food materials must undergo a considerable amount of structural changes during drying. The changes in structure may extensively vary depending on the type of drying. Through the manipulation of process parameters and drying technique, the structural deformation can be controlled to a certain degree [28–30]. Therefore, it is imperative to select a sustainable drying technique that effectuates minimum structural changes.

2.1.1.2 Porosity

Porosity is an essential quality attribute that delineates the extent to which a food tissue shrinks during drying [31]. It is defined as the empty spaces in the total volume of a food material [32]. The porosity and rehydration capacity considerably vary in different drying techniques [25, 33]. These two physical attributes have substantial effects on the quality of the final dried food [34]. Porosity has an inversely proportional relationship with bulk density.

2.1.1.3 Texture (Chewiness)

The texture of food materials relies mostly on physical properties, and it significantly influences the perception of the consumers [35–37]. Porosity substantially affects the characterization of the texture of foodstuffs [38]. Understanding the relationship between process parameters and dried food quality is imperative to attain a satisfactory texture [34]. The breakdown and flavor release of dried food in the mouth is impacted by the viscosity and food structure [38].

The textural properties of food materials are deeply affected by the water content. Intermediate to higher water content leads to juicy, soft, tender, chewy, and moist products; conversely, low water content food leads to crunchy, hard and crisp products (See Fig. 2.4) [39].

In the falling rate of drying, the surface may cripple if the energy supply is constant. Because, at this stage, a considerable amount of moisture is reduced, which may subsequently lead to the case hardening. Therefore, drying techniques should be selected in a way that the texture of the food materials remains intact, providing more importance on the crispness. Crispy food generally has a water percentage below 10% [40].

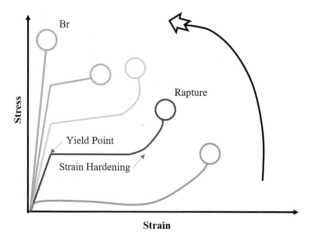

Fig. 2.4 Water and mechanical properties

2.1.1.4 Rehydration Properties

At a given time and temperature of the water, the original fresh weight gained by the dried samples during rehydration is defined as rehydration efficiency. The method of drying and the subsequent processing conditions are the key factors influencing rehydration efficiency. However, the porosity of dried product also has some effect on the rehydration properties [41].

2.1.2 Optical Properties

2.1.2.1 Color

Color is considered as a dominant quality attribute, which might change owing to various biochemical and chemical reactions during drying [42–44]. Color change reactions are studied following the principle of zero-order reaction kinetics [45]. In most plant tissues, a naturally occurring enzyme, known as phenolics, is present. Due to the enzymatic reaction during drying, brownish pigments are produced in the plant tissues. This phenomenon is observed as the enzymatic reaction essentially oxidizes the phenolics. Ascorbic acid browning, Maillard reaction, and caramelization are some of the other chemical reactions that may significantly affect the color of dried food materials.

2.1.2.2 Consumer Appeal

The appearance of dried food materials cannot be defined by a single word since it is a combination of food's shape, color, and texture. Dried foods are more likely to appeal to the consumers if it looks closer to the fresh ones [46]. Throughout drying, the color of food material will change, and the structure of food tissue pointedly influences the features of color change [47–49]. Moreover, due to shrinkage and browning reactions, there might be implausible changes in the color of dried products that may substantially affect the consumers' choice [50]. Hence, depending on the requirement of consumers, it is imperative to modify such aspects of drying.

2.1.3   Sensory Properties

Depending on the process parameters of drying, the sensory properties of dried food materials may vary. Physio-chemical changes in the pores and food matrix considerably affect the flavor (Sensory property) during drying [51, 52].

2.1.3.1   Aroma

In order to define the food quality, the flavor is considered as one of the most influential parameters. In fruits, over 200 volatile aroma constituents are present, among which only a single group is responsible for producing the specific food smell. During drying, the aroma constituents present in the food materials are promptly evaporated due to their highly volatile characteristics. Therefore, in order to reduce the loss of aroma constituents, it is essential to optimize the process parameters of drying. Moreover, retention of dried foods flavor can be achieved through proper storage.

2.1.3.2   Taste

Dried food would be more popular to consumers if the taste of dried food is satisfactory. Although it is a relative factor varying between person to person, the taste is an essential sensory property of dried food materials.

2.1.4   Biological Quality Attributes

2.1.4.1   Microbial Load

Reproduction of the microorganism inside the food material depends entirely on the water activity. There is a specific range of water activity that will resist the growth of microorganisms. Drying helps to reduce the water activity in food materials that subsequently hinders the microorganism growth. Two utmost vital points of water activities that can prevent the development of any pathogenic bacteria and microorganisms are 0.84 and 0.6, respectively. Due to high water activity, the shelf life of food materials may diminish promptly. The shelf life of most food materials may increase approximately 2–3 times if the water activity is reduced by a factor of 0.1 [53, 54] (See Fig. 2.5).

Figure 2.5 revealed that there is a small chance of microbial growth if the value of water activity is less than 0.6. Moreover, the chemical properties are directly influenced by the variation of water activity, as shown in Fig. 2.6.

Therefore, a sustainable drying technique, along with the process parameters, should be selected carefully in order to reduce the water activity and in the process, limiting the microorganism growth. Along with the depletion of microorganism growth, reduction of water activity through drying can offer a lot of positive output, as shown in Fig. 2.7.

2.1.5 Nutritional Quality Attributes

2.1.5.1 Retention of Nutrients

The nutritional qualities of food materials undoubtedly depend on the structure of the food tissue [55, 56]. The bio-accessibility of food nutrients is significantly influenced by the processing conditions, environmental conditions, and tissue structure [57, 58]. Since maximum nutrients are connected to the cellular matrix or persist inside cells, the bio-accessibility of nutrients may be hindered by the complexity of food structure [58, 59].

Nutrients are mostly present in cells and cell walls [57]; therefore, the porous characteristic of food materials, whether processed or fresh, permit the nutrients to transfer with minimum resistance. Additionally, exposure to oxygen and light, along with porosity and storage temperature are the factors that also influence the degradation of nutrient quality [60]. Inferior dried product with higher nutritional loss may result from over-drying [61]. Therefore, to attain the desired dried food qualities, it is indispensable to establish an optimum combination of the various process parameters [62].

2.2 Energy and Time

Among the major industrial processes, drying is undoubtedly the utmost energy-intensive technique that consumes almost 15% of all industrial energy usage [63, 64]. Improving the energy efficiency of drying even by 1% in a drying industry could result in as much as a 10% increase in profit [63]. Therefore, even a small improvement in energy efficiency in drying technology may lead to effective, sustainable global energy preservation. This is because of the removal of moisture from food materials necessitates a substantial amount of energy [65]. The energy is supplied in the form of heat throughout different processing stages. For instance, when

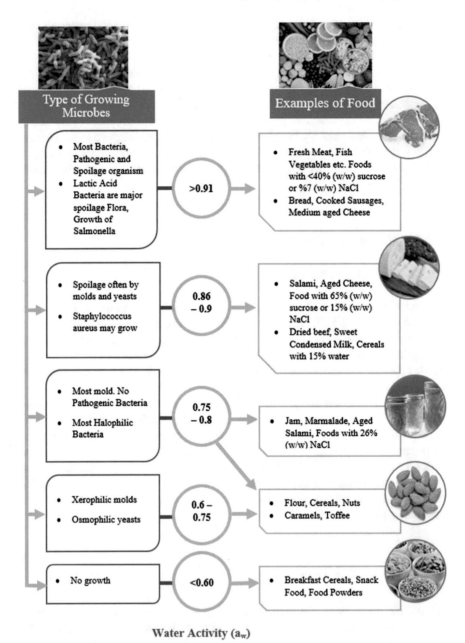

Water Activity (a_w)

Fig. 2.5 Water activity limit for growth of microorganisms of food and examples of foods with water activities over the range of various growth limits

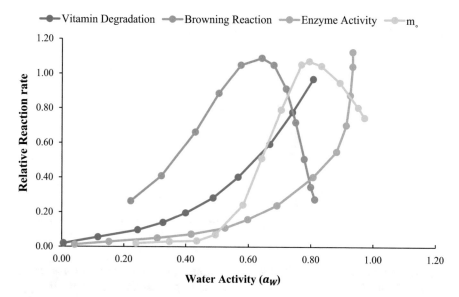

Fig. 2.6 Variation of stability with a water activity of food

Fig. 2.7 Positive outcome of reducing water activity through drying

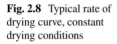

Fig. 2.8 Typical rate of drying curve, constant drying conditions

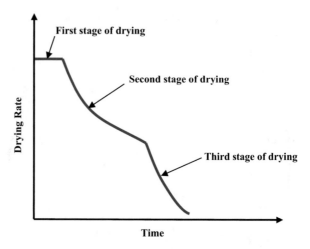

drying is accomplished at 50 °C, the required energy for moisture removal is almost 1.5×10^6 kJ/t [66]. However, with the variation of process parameters and the type of food sample, the required energy may vary [67–69].

The energy requirement for different steps of drying is different because of the complex phenomenon of moisture removal. A basic drying curve of food material is presented in Fig. 2.8, and the various stages of drying are explained below:

- At the first stage of drying, a constant drying rate is observed where the free water from the surface is removed [70].
- At the second stage of drying, which is also known as the first falling rate of drying, capillary water, and loosely bound water migrates to the surface. The drying rate of this stage depends on the internal parameters such as shape, size, and collapse of internal tissues [71, 72]. Due to the low moisture content, this phase is susceptible to heat damage [73].

- Finally, at the third stage, known as the second falling rate of drying, mainly the bound water is removed. This phase of drying requires a higher amount of energy compared to the other two phases due to the low thermal conductivity of the solid matrix of food materials [74]. Evacuation of the first 90% of the water from the food requires nearly as much energy as the final 10% of moisture content during drying [75]. A summary of the energy and time needed to dry some selected food materials is represented in Table 2.1 [76, 77].

From Table 2.1, it is evident that different food materials require a dissimilar amount of energy and time to dry because of the variation in the distribution of water as well as material properties [66]. Among the water content, the removal of bound water necessitates the maximum amount of energy; albeit, amount the bound water varies with the types of food [78]. Therefore, energy-efficient bound water removal from the food material will undoubtedly contribute to the energy sector of the developing countries.

Table 2.1 Energy and time requirement to dry some selected food materials

Sliced crop	Initial moisture Content, (%wb)	Final moisture Content, (%wb)	Energy required (MJ/kg)	Maximum temperature (°C)	Required drying temperature (°C)	Required drying time (hour)
Apple	80–85	20–24	23–50	70	45	8–10
Banana	70–80	7–15	16.79	70	45	8–10
Cassava chips	62–75	7–17	23–62	150	30–60	7–10
Pepper	75–80	5–14	16.1	90	40	5–6
Mango	80–85	12–18	15.64	70	55	10–15
Potato	70–75	8–13	14.53	75–85	50–70	10–13
Azarole	65–70	8–9	27–42	75	60–70	50

It is admirable to remark that the time and cost required for water removal is approximately proportional to the energy requirement. Furthermore, Greenhouse Gas emission from the drying techniques must be taken into consideration before removing an additional bound water from food samples.

2.3 Cost and Safety

Achieving the required milestones for the fundamental dimensions of sustainable development – social, environmental, and economic has become challenging for the world [79]. Despite the best efforts of the researchers, nearly one billion people are still prone to extreme poverty, and this situation is exacerbated due to the rise of income inequality among and within many developing countries. Additionally, inefficient production processes and unsustainable consumption behavior have racked up staggering social and economic losses. Hence, it is imperative to ensure proper investment in the development of new drying technologies. In order to achieve the benchmark for sustainable development, collaborative global actions are required for research and development while ensuring the seamless sharing of data. Thus, the outstanding cost of technological advancement can be minimized. The issues that might arise for adopting new drying technologies are discussed below.

When discussing the adoption capacity of a developing country, its supply of natural resources should be taken into consideration. A developing country stymied by an inadequate reserve of natural resources. Infertile land will undoubtedly face more difficulties in adopting advanced industrial drying techniques in contrast to a developed nation endowed with such resources. However, with the help of renewable energy and easily accessible technologies, these bounds of adaptivity can be reduced to a certain extent. For example, there are many solar drying systems developed, which can significantly reduce the dependence on traditional energy sources [80–82]. When farms collaborate as a community to develop sustainable drying facilities, it is much easier to utilize the advantages of improved drying

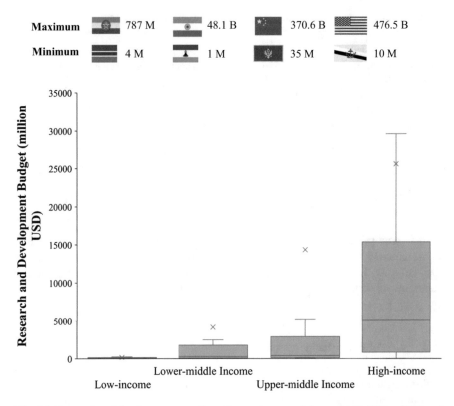

Fig. 2.9 Research and development spending of countries around the world (B: Billion USD and M: Million USD)

technologies. Fragmented or individual efforts may prove to be too costly and difficult to maintain. The investment facilities undertaken by various countries have been described in Fig. 2.9 [83]. It is apparent from the figure that despite higher GDP and income status, some of the higher-income countries have less budget for research and development than the low-income countries. This condition can be related to their ability to purchase exported goods from low-income and middle-income countries. Since many high-income countries can import required products (e.g. stored food) from other developing countries, they do not find the necessity to invest an excessive amount of funds for research in these areas. However, it is essential to note that developing countries are opting to spend a significant amount of research and development in order to improve their production process. This is a clear indication for their capacity to adopt novel and efficient drying process.

Essential services, primarily communication, ease of transportation, and infrastructure, are paramount for sustainable development. Developing countries with adequate bridges, railways, roads, and harbors can effectively transport the finished food products, which will reduce the food-processing costs in successive stages. As the food processing cost depletes, the additional capital can be invested to acquire novel drying technologies. Economy development is hindered in the absence of dependable infrastructure. It has been observed that developing countries invest for short term profits such as spending capital on a new steel mill, instead of automating the telephone system or digitalizing industries. As a result, the food export industry will lack behind. Therefore, systematic development of the entire industrial management facilities should be given maximum priority as opposed to standalone supply chain development.

An analysis of history shows that the economy suffering from interventions performed poorly in the 1950s, 1960s, and 1970s [84]. Developing countries like Myanmar, Argentina, Ghana, Tanzania, and Ethiopia had an interventionist economy that grew slowly, if at all. Ethiopia adopted rigid Soviet-style policies under its new government overthrowing their emperor. Their successive attempts to collectivize agriculture resulted in widespread famine. Some other developing countries, like Nigeria, Myanmar, and Ghana, were endowed with a strong economy right after they received their independence; however, the GDPs and living standards of those countries shrunk over time due to political unrest and insufficient infrastructure development. Although developing countries, such as Bangladesh, Kenya, and India, had a reasonably sustainable government than their more highly interventionist neighbors. The development is still disappointingly slow due to their propensity to conventional technologies [84].

Every country is endowed with a limited amount of naturals resources; therefore, it is crucial to invest said assets conservatively. The policymakers of the developing countries use numerous planning tools in order to transfer and develop environmentally sound technologies (ESTs), and one such tool is called Technology needs assessments (TNAs). The following section discusses the methods of proposing appropriate drying technologies for developing countries based on a critical evaluation of the TNAs report.

2.4 Method of Proposing Drying Technology for Developing Countries

A survey was conducted by the United Nations Framework Convention on Climate Change (UNFCC) in 31 developing countries to analyze their tendency to adopt new technologies based on various factors. Figure 2.10 shows that developing countries have an expected propensity to adopt new technologies in the agricultural sector [85]. This is indicative of the high percentage of investment and government subsidies, which are available for new technology adaption in developing countries.

Moreover, with a satisfactory yearly budget, most developing countries can equip state-of-the-art drying technologies for better performance.

Different types of drying technology use different modes of energy, which is essential for proposing drying technology to a developing country. Some developing countries may have higher solar energy potential, whereas some might have

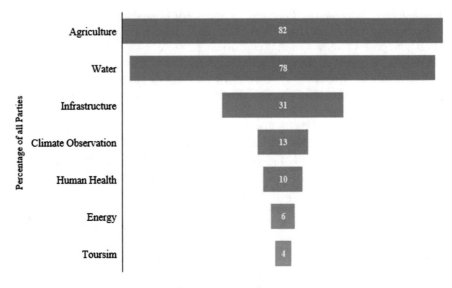

Fig. 2.10 Prioritized sector for adaption of new technologies

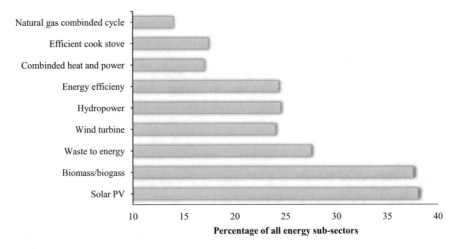

Fig. 2.11 Prioritized technologies for the energy industries

better geothermal energy [86, 87]. Figure 2.11 describes a relative comparison between the energy industries subsector where new technology integration is prioritized. Figure 2.11 concludes that most preferred sectors of energy industries are solar photovoltaic and biomass/biogas, followed by wind, hydropower, and geothermal. Thus, drying technologies that can incorporate solar energy and biomass energy can be considered as highly recommended for developing countries.

Beyond the energy requirements, the possibility of adopting new technologies is limited by numerous other development and logistical factors. Figure 2.12 presents

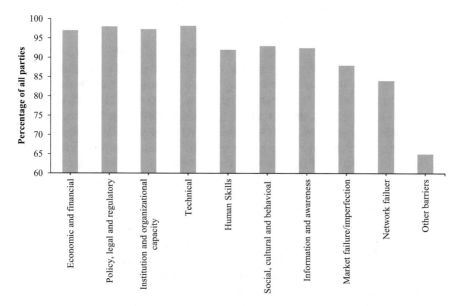

Fig. 2.12 Reported barriers to the development and transfer of technologies for adaption

which sectors act as barriers to the development and transfer of new technology adaption [85]. The figure shows that economic and financial challenges are the biggest barrier to the use of new technologies in developing countries, which is followed by policy, legal, and regulatory restriction. Hence, while proposing sustainable drying techniques, the economic and government regulations of respective developing countries should be considered.

Furthermore, Fig. 2.12 also highlights institutional, organizational capacity, technical, human skills, information, and awareness as significant challenges for deploying new technologies in the existing infrastructure. It would be difficult to propose some of the advanced drying technologies for developing countries despite their high efficiency and low energy requirement since the developing countries will not be able to provide the required human skill and technical support for proper maintenance. Lack of awareness will also become challenging to create market values for someone for the newest technologies.

Despite the restrictions, more and more countries are moving forward and accepting new technologies to improve the status of their economy. Various free market has opened the possibility for seamless transition of ideas and technologies.

Additionally, the entrepreneur mindset of the new millennials has created a ground for mass investments behind new technologies. In recent years developing countries like Azerbaijan and Kenya have spent 2.17 million USD and 29.38 million USD, respectively, for a technology action plan (TAP) [85]. Whereas in South Asian developing countries, such as Bangladesh has spent an estimated 6.25 million USD in agricultural research and Sri Lanka spending 28.42 million USD for the transportation system [85].

Drying, albeit a ubiquitous technique, most developing countries use the conventional drying approach, which is simple in operation but energy inefficient. Despite their conceivable limitation, the existing drying techniques used in developing countries have excellent market penetration. From the root level farmers to the finished product distributor, every stage of the food supply chain is accustomed to these existing technologies. In order to understand the requirement of the developing countries and subsequently propose an improved drying approach, a discussion on the positive features of the existing drying technique is essential. This will highlight the manufacture's reservation towards adopting advanced drying technologies.

The following chapter (Chap. 3) of this book will highlight the energy condition and available resources present in low-income and lower-middle-income developing countries. Chapter 4 will discuss the drying technologies currently practiced in developing countries, their applications, and their limitations. Afterward, Chap. 5 will propose novel drying technologies with improved efficiency. These proposed techniques were assessed in order to evaluate their potential application. Finally, the challenges that might occur during the adoption of recommended technologies are presented critically in Chap. 6.

References

1. M. Masud, M.U.H. Joardder, M.T. Islam, M.M. Hasan, M.M. Ahmed, Feasibility of utilizing waste heat in drying of plant-based food materials, in *International Conference on Mechanical, Industrial and Materials Engineering*, RUET, Rajshahi, Bangladesh, 500 (2017)
2. T. Koyuncu, Y. Pinar, F. Lule, Convective drying characteristics of azarole red (Crataegus monogyna Jacq.) and yellow (Crataegus aronia Bosc.) fruits. J. Food Eng. **78**(4), 1471–1475 (2007)
3. M.A. Karim, M.N.A. Hawlader, Drying characteristics of banana: Theoretical modelling and experimental validation. J. Food Eng. **70**(1), 35–45 (2005)
4. M.A. Karim, M.N.A. Hawlader, Mathematical modelling and experimental investigation of tropical fruits drying. Int. J. Heat Mass Transf. **48**(23), 4914–4925 (2005)
5. M.U.H. Joardder, M.H. Masud, S. Nasif, J.A. Plabon, S.H. Chaklader, Development and performance test of an innovative solar derived intermittent microwave convective food dryer, in *AIP Conference Proceedings*, **2121**(1), 40010–40013 (2019)
6. M.U.H. Joardder, M. Mourshed, M.H. Masud, Characteristics of bound water, in *State of Bound Water: Measurement and Significance in Food Processing*, ed. Springer International Publishing, Cham, 29–45 (2019)
7. J.C. Ho, S.K. Chou, K.J. Chua, A.S. Mujumdar, M.N.A. Hawlader, Analytical study of cyclic temperature drying: Effect on drying kinetics and product quality. J. Food Eng. **51**(1), 65–75 (2002)
8. E. J. Quirijns, *Modelling and Dynamic Optimisation of Quality Indicator Profiles during Drying, Wageningen University* (2006)
9. M.U. Joardder, M.H. Masud, M.H. Azharul, Relationship between intermittency of drying, microstructural changes, and food quality. Intermittent and nonstationary drying technologies: Principles and applications, 123 (2017)

10. S.J. Kowalski, A. Pawłowski, Energy consumption and quality aspect by intermittent drying. Chem. Eng. Process. Process Intensif. **50**(4), 384–390 (2011)

11. M.U.H. Joardder, M.H. Masud, Effectiveness of food preservation systems, in *Food Preservation in Developing Countries: Challenges and Solutions*, Springer, 127–152 (2019)

12. N. Duc Pham et al., Quality of plant-based food materials and its prediction during intermittent drying. Crit. Rev. Food Sci. Nutr. **59**(8), 1197–1211 (2019)

13. M.S. Rahman, O.J. McCarthy, A classification of food properties. Int. J. Food Prop. **2**(2), 93–99 (1999)

14. M.U.H. Joardder, M.H. Masud, A brief history of food preservation, in *Food Preservation in Developing Countries: Challenges and Solutions*, Springer, 57–66 (2019)

15. M.M. Rahman, M.U.H. Joardder, A. Karim, Non-destructive investigation of cellular level moisture distribution and morphological changes during drying of a plant-based food material. Biosyst. Eng. **169**, 126–138 (2018)

16. M.M. Rahman, Y.T. Gu, M.A. Karim, Development of realistic food microstructure considering the structural heterogeneity of cells and intercellular space. Food Struct. **15**, 9–16 (2018)

17. M.U.H. Joardder, C. Kumar, M.A. Karim, Food structure: Its formation and relationships with other properties. Crit. Rev. Food Sci. Nutr. **57**(6), 1190–1205 (2017)

18. A.K. Datta, Porous media approaches to studying simultaneous heat and mass transfer in food processes. I: Problem formulations. J. Food Eng. **80**(1), 80–95 (2007)

19. A.K. Datta, Porous media approaches to studying simultaneous heat and mass transfer in food processes. II: Property data and representative results. J. Food Eng. **80**(1), 96–110 (2007)

20. J. Srikiatden, J.S. Roberts, Moisture transfer in solid food materials: A review of mechanisms, models, and measurements. Int. J. Food Prop. **10**(4), 739–777 (2007)

21. M.U.H. Joardder, M. Mourshed, M.H. Masud, Challenges in bound water measurement, in *State of Bound Water: Measurement and Significance in Food Processing*, ed., Springer International Publishing, Cham, 83–92 (2019)

22. A. Halder, A.K. Datta, R.M. Spanswick, Water transport in cellular tissues during thermal processing. AICHE J. **57**(9), 2574–2588 (2011)

23. S.W. Fanta et al., Microscale modeling of coupled water transport and mechanical deformation of fruit tissue during dehydration. J. Food Eng. **124**, 86–96 (2014)

24. M.I.H. Khan, R.M. Wellard, S.A. Nagy, M.U.H. Joardder, M.A. Karim, Investigation of bound and free water in plant-based food material using NMR T2 relaxometry. Innov. Food Sci. Emerg. Technol. **38**, 252–261 (2016)

25. M.U.H. Joardder, A. Karim, C. Kumar, Effect of temperature distribution on predicting quality of microwave dehydrated food. J. Mech. Eng. Sci. **5**, 562–568 (2013)

26. M.U.H. Joardder, M.H. Masud, Challenges and mistakes in food preservation, in *Food Preservation in Developing Countries: Challenges and Solutions*, ed., Springer International Publishing, Cham, 175–198 (2019)

27. M. Karel, D.B. Lund, *Physical Principles of Food Preservation: Revised and Expanded*, 129, CRC Press (2003)

28. F. Kong, R.P. Singh, Chemical deterioration and physical instability of foods and beverages, in *The Stability and Shelf Life of Food*, Second edn., Elsevier, 43–76 (2016)

29. S.H. Anwar, B. Kunz, The influence of drying methods on the stabilization of fish oil microcapsules: Comparison of spray granulation, spray drying, and freeze drying. J. Food Eng. **105**(2), 367–378 (2011)

30. M.A.M. Khraisheh, Y.S. Al-degs, W.A.M. Mcminn, Remediation of wastewater containing heavy metals using raw and modified diatomite. Chem. Eng. J. **99**(2), 177–184 (2004)

31. A.N.A.M.I.B. Ayrosa, R.N.D.E.M. Pitombo, Influence of plate temperature and mode of rehydration on textural parameters of precooked freeze-dried beef. J. Food Process. Preserv. **27**(3), 173–180 (2003)

32. M.S. Rahman, Food stability beyond water activity and glass transition: Macro-micro region concept in the state diagram. Int. J. Food Prop. **12**(4), 726–740 (2009)

33. M.U.H. Joardder, A. Karim, C. Kumar, R.J. Brown, Determination of effective moisture diffusivity of banana using thermogravimetric analysis. Procedia Eng. **90**, 538–543 (2014)
34. M. U. H. Joardder, A. Karim, C. Kumar, R. J. Brown, *Porosity: Establishing the Relationship between Drying Parameters and Dried Food Quality, Springer* (2015)
35. P.C. Moyano, E. Troncoso, F. Pedreschi, Modeling texture kinetics during thermal processing of potato products. J. Food Sci. **72**(2), E102–E107 (2007)
36. N. Wang, J.G. Brennan, Changes in structure, density and porosity of potato during dehydration. J. Food Eng. **24**(1), 61–76 (1995)
37. C. Wilkinson, G.B. Dijksterhuis, M. Minekus, From food structure to texture. Trends Food Sci. Technol. **11**(12), 442–450 (2000)
38. H. Schubert, Food particle technology. Part I: Properties of particles and particulate food systems. J. Food Eng. **6**(1), 1–32 (1987)
39. J. Blahovec, Role of water content in food and product texture. Int. Agrophysics **21**(3), 209 (2007)
40. G. Roudaut, C. Dacremont, B.V. Pàmies, B. Colas, M. Le Meste, Crispness: A critical review on sensory and material science approaches. Trends Food Sci. Technol. **13**(6–7), 217–227 (2002)
41. P.P. Lewicki, Some remarks on rehydration of dried foods. J. Food Eng. **36**(1), 81–87 (1998)
42. P.P. Lewicki, E. Duszczyk, Color change of selected vegetables during convective air drying. Int. J. Food Prop. **1**(3), 263–273 (1998)
43. O. Boeh-Ocansey, Effects of vacuum and atmospheric freeze-drying on quality of shrimp, turkey flesh and carrot samples. J. Food Sci. **49**(6), 1457–1461 (1984)
44. K.J. Chua, A.S. Mujumdar, S.K. Chou, M.N.A. Hawlader, J.C. Ho, Convective drying of banana, guava and potato pieces: Effect of cyclical variations of air temperature on drying kinetics and color change. Dry. Technol. **18**(4–5), 907–936 (2000)
45. A. Maskan, S. Kaya, M. Maskan, Effect of concentration and drying processes on color change of grape juice and leather (pestil). J. Food Eng. **54**(1), 75–80 (2002)
46. V.P. Oikonomopoulou, M.K. Krokida, Novel aspects of formation of food structure during drying. Dry. Technol. **31**(9), 990–1007 (2013)
47. A. Reyes, P.I. Alvarez, F.H. Marquardt, Drying of carrots in a fluidized bed. I. Effects of drying conditions and modelling. Dry. Technol. **20**(7), 1463–1483 (2002)
48. S. Grabowski, M. Marcotte, M. Poirier, T. Kudra, Drying characteristics of osmotically pretreated cranberries—Energy and quality aspects. Dry. Technol. **20**(10), 1989–2004 (2002)
49. C.F. Hansmann, E. Joubert, T.J. Britz, Dehydration of peaches without sulphur dioxide. Dry. Technol. **16**(1–2), 101–121 (1998)
50. M. Mahiuddin, M.I.H. Khan, C. Kumar, M.M. Rahman, M.A. Karim, Shrinkage of food materials during drying: Current status and challenges. Compr. Rev. Food Sci. Food Saf. **17**(5), 1113–1126 (2018)
51. C. Druaux, A. Voilley, Effect of food composition and microstructure on volatile flavour release. Trends Food Sci. Technol. **8**(11), 364–368 (1997)
52. C. Lafarge, M. Bard, A. Breuvart, J. Doublier, N. Cayot, Influence of the structure of cornstarch dispersions on kinetics of aroma release. J. Food Sci. **73**(2), S104–S109 (2008)
53. R.K. Singh, D.B. Lund, Kinetics of ascorbic acid degradation in stored intermediate moisture apples, in *Proceedings of the 3rd International Congress on Engineering and Food. Engineering Sciences in the Food Industry*, vol. 1, (1984)
54. M.C. Vieira, A.A. Teixeira, C.L.M. Silva, Kinetic parameters estimation for ascorbic acid degradation in fruit nectar using the partial equivalent isothermal exposures (PEIE) method under non-isothermal continuous heating conditions. Biotechnol. Prog. **17**(1), 175–181 (2001)
55. S. Palzer, Food structures for nutrition, health and wellness. Trends Food Sci. Technol. **20**(5), 194–200 (2009)
56. R. Sharma, Food structures and delivery of nutrients. *Food Mater. Sci. Eng.*, 204–221 (2012)
57. J. Parada, J.M. Aguilera, Food microstructure affects the bioavailability of several nutrients. *J. Food Sci.* **72**(2) (2007)

58. I. Sensoy, A review on the relationship between food structure, processing, and bioavailability. Crit. Rev. Food Sci. Nutr. **54**(7), 902–909 (2014)

59. K.W. Waldron, M.L. Parker, A.C. Smith, Plant cell walls and food quality. Compr. Rev. Food Sci. Food Saf. **2**(4), 128–146 (2003)

60. S.S. Sablani, Drying of fruits and vegetables: Retention of nutritional/functional quality. Dry. Technol. **24**(2), 123–135 (2006)

61. K. Franzen, R.K. Singh, M.R. Okos, Kinetics of nonenzymatic browning in dried skim milk. J. Food Eng. **11**(3), 225–239 (1990)

62. S. Ghnimi, S. Umer, A. Karim, A. Kamal-Eldin, Date fruit (Phoenix dactylifera L.): An underutilized food seeking industrial valorization. NFS J. **6**, 1–10 (2017)

63. S.K. Chou, K.J. Chua, New hybrid drying technologies for heat sensitive foodstuffs. Trends Food Sci. Technol. **12**(10), 359–369 (2001)

64. C. Kumar, M.U.H. Joardder, T.W. Farrell, G.J. Millar, M.A. Karim, Mathematical model for intermittent microwave convective drying of food materials. Dry. Technol. **34**(8), 962–973 (2016)

65. C. Kumar, M.A. Karim, M.U.H. Joardder, Intermittent drying of food products: A critical review. J. Food Eng. **121**, 48–57 (2014)

66. N.R. Nwakuba, S.N. Asoegwu, K.N. Nwaigwe, Energy requirements for drying of sliced agricultural products: A review. Agric. Eng. Int. CIGR J. **18**(2), 144–155 (2016)

67. G.S.V. Raghavan, T.J. Rennie, P.S. Sunjka, V. Orsat, W. Phaphuangwittayakul, P. Terdtoon, Overview of new techniques for drying biological materials with emphasis on energy aspects. Brazilian J. Chem. Eng. **22**(2), 195–201 (2005)

68. M.A. Billiris, T.J. Siebenmorgen, A. Mauromoustakos, Estimating the theoretical energy required to dry rice. J. Food Eng. **107**(2), 253–261 (2011)

69. S. Gunasekaran, T.L. Thompson, Optimal energy management in grain drying. Crit. Rev. Food Sci. Nutr. **25**(1), 1–48 (1986)

70. A.S. Mujumdar, A.S. Menon, Drying of solids: Principles, classification, and selection of dryers. Handb. Ind. Dry. **1**, 1–39 (1995)

71. M.I.H. Khan, S.A. Nagy, M.A. Karim, Transport of cellular water during drying: An understanding of cell rupturing mechanism in apple tissue. Food Res. Int. **105**, 772–781 (2018)

72. M.I.H. Khan, T. Farrell, S.A. Nagy, M.A. Karim, Fundamental understanding of cellular water transport process in bio-food material during drying. Sci. Rep. **8**(1), 15191 (2018)

73. S.J. Kowalski, A. Pawłowski, Modeling of kinetics in stationary and intermittent drying. Dry. Technol. **28**(8), 1023–1031 (2010)

74. A. Motevali, S. Minaei, A. Banakar, B. Ghobadian, H. Darvishi, Energy analyses and drying kinetics of chamomile leaves in microwave-convective dryer. J. Saudi Soc. Agric. Sci. **15**(2), 179–187 (2016)

75. M.U.H. Joardder, R.J. Brown, C. Kumar, M.A. Karim, Effect of Cell Wall properties on porosity and shrinkage of dried apple. Int. J. Food Prop. **18**(10), 2327–2337 (2015)

76. M.U.H. Joardder, M. Mourshed, M.H. Masud, Water in foods BT, in *State of Bound Water: Measurement and Significance in Food Processing*, ed. Springer International Publishing, Cham, 7–27 (2019)

77. W. Weisis, J. Buchinger, Solar drying: establishment of a production, sales and consulting infrastructure for solar thermal plants in Zimbabwe. Arbeitsgemeinschaft Erneuerbare Energie (AEE) of the Institute for Sustainable Technologies, Austria (2003)

78. M.R. Okos, Food dehydration, in *Hand Book Food Engineering* (1992)

79. U. Nations, *World Economic and Social Survey 2013 Sustainable Development Challenges* (2013)

80. M.H. Masud, R. Ahamed, M. Mourshed, M.Y. Hossan, M.A. Hossain, Development and performance test of a low-cost hybrid solar air heater. *Int. J. Ambient Energy* **40**(1) (2019)

81. M.A. Karim, M. Hawlader, Development of solar air collectors for drying applications. Energy Convers. Manag. **45**(3), 329–344 (2004)

82. M.A. Karim, M.N.A. Hawlader, Performance investigation of flat plate, v-corrugated and finned air collectors. Energy **31**(4), 452–470 (2006)
83. J. Desjardins, Infographic_ Visualizing How Much Countries Spend on R&D, 2018. [Online]. Available: https://www.visualcapitalist.com/money-country-puts-r-d/. Accessed: 08-Aug-2019
84. I. Y. Naturefriends, Sustainable Development and its Challenges in Developing Countries, 2016. [Online]. Available: http://www.iynf.org/2018/08/a-guide-to-sustainable-development-and-its-challenges-in-developing-countries/. Accessed: 08-Aug-2019
85. U. Nation, United Nations Framework Convention on Climate Change, What are the Technology Needs of Developing Countries? (2014)
86. N.N. Mustafi, M. Mourshed, M.H. Masud, M.S. Hossain, M.R. Kamal, Feasibility test on green energy harvesting from physical exercise devices, in *AIP Conference Proceedings*, **1851**, (2017)
87. S.K. Ghosh, M. Mourshed, P.C. Karmaker, A. Ahmed, M.H. Masud, Performance enhancement of solar PV/T by using glycerin as an additive, in *AIP Conference Proceedings*, **2121**(1), 130002-130005 (2019)

Chapter 3
Energy and Drying

Energy sustainability is critical for this world because every system from human to machine requires energy to be active. The source of energy in this world can be divided into two main groups: (i) conventional fossil fuel (coal, oil, natural gas) that is limited and (ii) renewable energy (solar, wind, hydro, geothermal, etc.) which is abundant. Researches on conventional fossil energy show an alarming concern that the fuel reserve will be almost finished at the end of this century. Therefore, people are finding alternative sustainable sources of energy, which can be replenished [1]. Despite having the significant renewable energy potential, the human being has failed to exploit the full potential of it [2]. The reasons for this failure are discussed in the renewable energy section of this chapter. In order to utilize the complete potential of renewable energy, it should not be used only for electricity generation; instead, renewable energy should be used in all sectors where energy is required. Using renewable energy for food drying can be one of the most viable options because existing industrial drying processes consume 20–25% of the total energy used in the food processing industry [3, 4]. Therefore, proper implementation of renewable energy technologies in the field of food drying can significantly reduce the energy requirement of food process industries and subsequently reduces the greenhouse gas (GHG) emission [5, 6]. This chapter comprehensively discusses both the conventional and renewable energy scenario of the world. Moreover, it has also provided an overview of energy requirements in drying and the barriers in renewable energy integration in food drying.

© Springer Nature Switzerland AG 2020
M. Hasan Masud et al., *Sustainable Food Drying Techniques
in Developing Countries: Prospects and Challenges*,
https://doi.org/10.1007/978-3-030-42476-3_3

3.1   Non-renewable Energy Usages Around the World

3.1.1   Coal

Coal is one of the oldest sources of conventional energy, which can be used for power generation and in other large-scale industries. However, this energy source is decreasing at an alarming rate. Coal is also responsible for substantial CO_2 emissions. Major share of coal is used in electricity production, and the leading coal consumers for electricity generation are India (75% of electricity is generated using coal) and Indonesia (58% of electricity is generated using coal)[7]. It is evident that higher coal user countries generally belong to the lower and lower-middle-income countries. In contrast, upper middle income and higher-income countries like the USA, China, and the EU countries consume a comparatively lower amount of coal, according to the International Energy Agency (IEA)[7].

Figure 3.1 presents the reserve of coal in different regions of the world in the years of 1998, 2008, and 2018 [8]. It shows that the total coal reserve was higher in 1998 (1,058,811 million tons) compared with 2008 (888,301 million tons), which indicates that coal reserve depleted considerably during this period. Unplanned and excessive consumption of resources in the industry level is the primary reason behind the unexpected reduction of coal reserves from 1998 to 2008. The comparison of global coal reserves between 2008 and 2018 shows that the amount of coal reserve is higher in 2018 (1,054,782 million Tons) than 2008 (888,301 million Tons). In the North America region and Asia Pacific region, due to numerous coal mines, discoveries, and planned consumption, the availability of coal has increased significantly. Although the net global coal reserve is increasing, a decrease in coal supply can be seen in North American region, Commonwealth of Independent

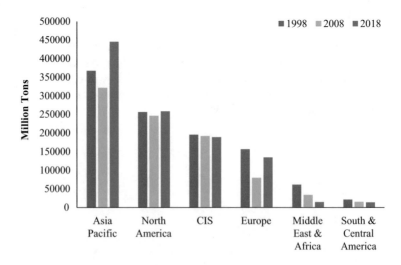

Fig. 3.1 Coal distribution around the globe in different years

Table 3.1 Proportion of electricity produced from oil in different countries by economic group

Economic groups	Countries	Electricity production from oil sources (% of total)	
		Year	
		2000	2015
Lower-income countries	Haiti	48.263	92
	Niger	34.47	57.62
	Tanzania	10.92	21.94
	Mozambique	0.433	0.763
Lower middle-income countries	Angola	36.87	46.83
	Bangladesh	6.47	16.38
	Egypt	13	21
	India	5.19	1.66
Upper middle-income countries	Belarus	6.57	1.06
	China	3.49	0.17
	Dominican Republic	90.85	53.48
	Gabon	20.53	10.21
Higher-income countries	Denmark	1.95	1
	Saudi Arabia	27.63	20.4
	Sweden	1.056	0.156
	United States of America	2.94	0.9

States (CIS) region, and in Middle East & Africa and South & Central America. This confirms a higher use of coal in these selected regions.

According to literature, the total coal reserve until December 2018 was 1,054,782 million tons, and the consumption rate of coal is currently 7.89 Billion tons per year [9, 10]. It is anticipated that the world will run out of coal in the next 150 years if the consumption rate exists at the current level [9]. The government and private sector should invest in renewable energy sources to meet the consumer energy demand.

3.1.2 Oil

Oil is one of the most essential energy sources that is used as transportation fuel, for heating and electricity generation, etc. The higher consumption rate of oil produces a substantial amount of greenhouse gas and exacerbates global warming that is shifting people towards renewable clean energy sources to meet their demand [11, 12]. Table 3.1 presents the yearly oil-based electricity production for some selected countries from different economic groups [13]. According to Table 3.1, for developing countries, consumer dependency on oil has increased significantly over the last 15 years (from 2000 to 2015). However, this consumption of oil has decreased for higher-income countries like Denmark, Saudi Arabia, the USA, etc. This is due to their financial and logistical ability to invest in renewable energy sources.

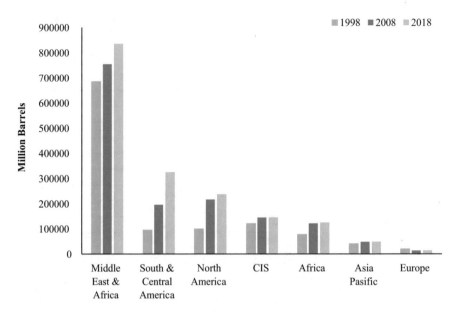

Fig. 3.2 Oil reserve by regions in the years 1998, 2008 and 2018 respectively

Lower-income and lower-middle-income countries do not have strong financial background to invest in renewable energy research and development programs in order to utilize the clean power resources. As such, countries in this income group moving to renewable energy sources slowly. Therefore, these countries depend on conventional energy sources like oil to meet their energy demand.

Figure 3.2 demonstrates the reserve of oil in different regions of the world in the years of 1998, 2008, and 2018 [][8]. It can be seen that the oil reserve is surprisingly increasing because of the detection of new oil mines by implementing advanced technology.

Oil demand in the Asia Pacific region has climbed from 26.04 Million Barrels Per Day (BPD) in 2007 to 34.57 million BPD in 2017 [14]. The average consumption rate of oil around the world is now about 99.84 million barrels per day [15], and it is estimated that the world will run out of oil in the next 53 years [16]. Therefore, more research should be done on the utilization of renewable energy sources in order to mediate the impending energy crisis.

3.1.3 Natural Gas

Another widely used conventional non-renewable energy source is natural gas. The global reserve of natural gas is also decreasing due to its excessive and unplanned use. The use of natural gas includes electricity production, industrial usage, household activity, commercial activity, transportation, etc. Among these, electricity

Table 3.2 Percentage of electricity produced from natural gas in some countries from different economic regions

Economic groups	Countries	Percentage of electricity generated from natural gas (%)	
		In year 2000	In year 2015
Lower-Income countries	Mozambique	0.021	12.83
	Yemen, Rep.	0	39.81
	Tanzania	0	43.91
	Congo, Dem. Rep.	0	46.65
Lower-Middle income countries	Bangladesh	88.78	80.70
	Bolivia	43.97	66.35
	Egypt	53.74	70.73
	Ghana	0	38.85
Higher-Middle income countries	Armenia	45.18	35.92
	Bulgaria	4.70	3.82
	Kazakhstan	10.68	18.39
Higher-Income countries	Australia	13.11	12.6
	Denmark	24.34	6.27
	Estonia	7	0.60
	Finland	14.48	7.57
	Argentina	54.65	49.48

production consumes major share of global natural gas stock. For example, in 2018 USA used about 301 billion cubic meters for electricity generation [17]. Table 3.2 presents the proportion of electricity produced from natural gas in some selected countries from different economic groups [13]. It shows that lower-income countries (e.g., Mozambique, Yemen Republic, Tanzania, the Democratic Republic of the Congo) and Lower- Middle-income countries (e.g., Bolivia, Egypt, Ghana) have an increasing trend of natural gas usages for the generation of electricity. On the other hand, Higher-Middle income countries like Armenia, Bulgaria, Kazakhstan, and Higher-income countries like Argentina, Australia, Denmark, Estonia, Finland are showing decreasing trend in the percentage of natural gas-based electricity production. Some countries in Lower-Middle income range like Bangladesh are also showing decreasing trend in percentage of natural gas-based electricity production.

Figure 3.3 presents the reserve of natural gas around different regions of the world in the years 1998, 2008, and 2018 [8]. From the figure, it is clear that the major reserve of natural gas is in the Middle East, followed by the Commonwealth of Independent States (CIS). The lowest reserve belongs to the South & Central America region. Figure 3.3 also shows that the world's total natural gas reserve has increased from the year 1998 to 2008. It can also be seen from Fig. 3.3 that in 2018, the natural gas reserve is largest than ever recorded, with 196,900 billion cubic meters. It was possible because of the discovery of new gas wells.

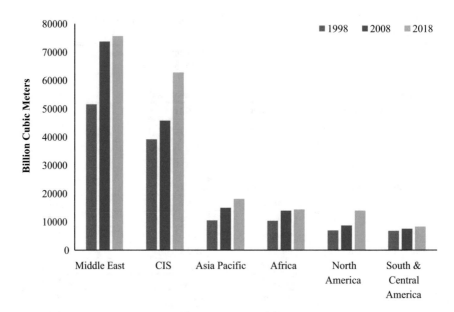

Fig. 3.3 Natural gas reserve of different regions around the world by different years

According to the Organization of Economic Co-operation & Development (OECD), at the end of December 2018, the production rate of natural gas was 130 billion cubic meters per year [18]. With the current consumption rate, it is estimated that the world will run out of its natural gas reserve in the next 51 years [8]. An overview of all the non-renewable energy (i.e. Coal, Oil, and Natural gas) reserves of selected countries from different income groups is presented in Table 3.3.

From Table 3.3, it can be seen that conventional fossil energy reserve is limited; hence, it is essential to reduce consumer dependency on these non-renewable energy sources. For instance, Canada is the third-largest contributor to world Oil supply, but the current consumption rate of oil 1679 million barrel/year will use up the entire oil reserve of Canada in 99.29 years. Canada being a developed country has the potential to find alternative energy sources; however, the consequences of running out of conventional fuel will be far more severe in developing countries. Figure 3.4 shows the lifetime of Coal, Oil, and Natural Gas reserves in some selected developing countries of the world according to the current global consumption rate.

Figure 3.4 shows that the natural gas of Bangladesh, Philippines, Congo, and Egypt will run out in 28, 25, 60, and 37 years, respectively. In the case of coal, the lowest lifetime is in Bangladesh (23 years) and highest in Philippines (93 years). Oil of Egypt, Congo, Niger, Philippines, and Bangladesh will run out in 14, 25, 59, 27, and 150 years respectively. Therefore, the information from the figure is alarming and shows the urgency of shifting towards renewable energy sources from conventional energy sources. Therefore, developing countries should invest in renewable energy-based technologies.

Table 3.3 Coal, Oil and Natural gas reserve of some selected countries from different economic group

Economic Regions	Countries	Coal (million tons)	Oil (million barrel)	Natural Gas (billion cu. m.)	References
Lower income countries	Democratic Republic of the Congo	87.9	2982	285	[19, 20]
	Ethiopia	297	2700	198.22	[21–23]
	Tajikistan	4300	12	5.66	[24, 25]
	Afghanistan	–	962	1472.47	[26]
	Chad	0.456	1500	999.5	[27, 28]
	Yemen	–	3000	478.55	[29, 30]
	Niger	178	150	18.6	[31, 32]
	Benin	–	8	1.133	[33, 34]
Lower middle-income countries	Angola	–	8160	383	[35]
	Bangladesh	2013	28	767.95	[36–38]
	Bolivia	0.99	465	280.33	[39–41]
	India	96468	4495	1330	[42, 43]
	Myanmar	543.75	105.78	637.13	[44, 45, 30]
	Pakistan	207	332	588.70	[46, 47]
	Philippines	2390	138.5	98.54	[48, 37]
	Nigeria	2000	36,972	5675	[49, 50]
	Congo	88	2982	285	[20, 51]
	Egypt	21	2400	1727.33	[52–54]
	Vietnam	3116	4400	699.43	[55, 56]
Upper middle-income countries	Algeria		12200	4504	[57, 58]
	China	130851	25000	5440	[59, 60]
	Colombia	5119.24	2000	133	[61]
	Ecuador	24	8273	11	[62, 63]
	Equatorial Guinea	0.382	1100	42	[64, 65]
	Gabon	–	2000	26	[66]
	Iraq	–	145019	3729	[67]
	Libya	–	48363	1505	[68]
	Malaysia	180	4000	1182.79	[69–71]
	Russian federation	69634	80000	47798.84	[72]
	Turkey	551	270	6.17	[73]

(continued)

Table 3.3 (continued)

Economic Regions	Countries	Coal (million tons)	Oil (million barrel)	Natural Gas (billion cu. m.)	References
Higher Income Countries	Australia	70927	3900	1989	[74, 75]
	Canada	4346	166700	1953.86	[76, 77]
	Japan	340	40	20900	[78–80]
	Kuwait	–	101500	1784	[81]
	United Arab Emirates	–	97800	6091	[82]
	Saudi Arabia	185000	267026	9069	[83, 84]
	New Zealand	825	1000	28.32	[85, 86]
	Qatar		25700	25000	[8, 87]
	United States of America	220167	293000	69631.12	[88–90]
	United Kingdom	29	20000	28316.84	[91, 92]

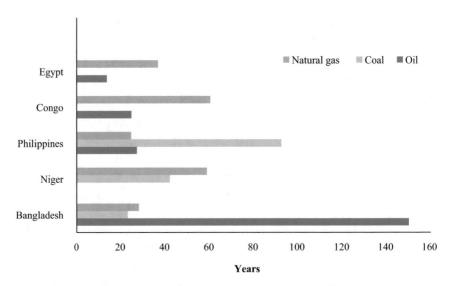

Fig. 3.4 Lifetime of conventional energy sources, i.e., Oil, Coal & Natural gas in years of some selected developing countries

3.2 Renewable Energy

Renewable energy, also known as Clean energy/ Green energy/ Sustainable energy/ Alternative Energy that is generated from natural processes, which are continuously replenished. It includes sunlight, wind, tidal, geothermal, and biomass energy. Figure 3.5 shows the different forms of renewable energy sources.

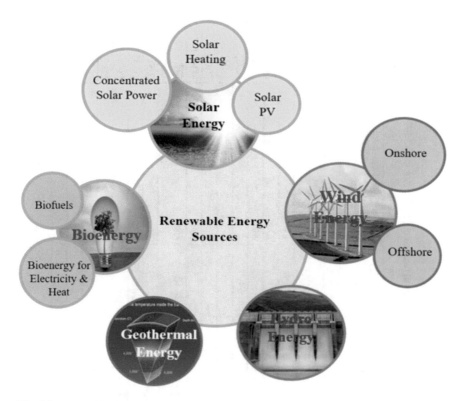

Fig. 3.5 Renewable energy sources

Renewable energy can be used in several sectors, and its application includes electricity generation, transportation, air conditioning, water pumping, industrial, household, and commercial use. However, renewable energy is mostly used for power generation. For example, in the year 2016, about 10% of total energy and 15% of total electricity in the USA is generated from renewable energy sources [93]. The benefit of renewable energy is that it emits almost no greenhouse gas and it is reusable. Because of these positive features, world is leaning towards renewable energy.

Table 3.4 shows the amount of electricity generation from various sources of renewable energy and the percentage share of renewable energy-based electricity generation for different lower and lower-middle-income countries. The table indicates that the percentage of renewable energy in total electricity generation for these countries is unsatisfactory, as most of the countries only generate 3–7% of its total electricity from renewable energy. However, some countries have a higher share of renewable energy contributions in generating electricity, such as Uganda (78.6%) [94], Kenya (70%) [95], Rwanda (54.4%) [96] and India (15.9%) [97]. This inspiring scenario of renewable energy-based electricity generation is indicative of the potential of developing countries to increase their clean power generation 5 to 6

Table 3.4 Electricity generation from renewable sources and percentage share of renewable energy in total electricity generation of different lower and lower-middle-income countries

Economic groups	Countries	Solar (MW)	Wind (MW)	Hydro (MW)	Geothermal (MW)	Bioenergy (MW)	Renewable share in total electricity generation(%)	References
Lower income countries	Ethiopia	14	324	2250	7.5	25	8	[98–100]
	Liberia	0.1	–	88	–	–	–	[101]
	Mali	4.5	–	250	–	–	–	[102]
	Rwanda	12.02	–	101.06	–	1.6	54.4	[96]
	Tanzania	7	–	602	100	200	4.9	[103]
	Uganda	50	–	400	–	17	78.6%	[104, 105]
Lower Middle-Income countries	Bangladesh	325.8	2.8	230	–	5.6	2.95	[106]
	Bolivia	60	–	658	–	–	–	[107]
	Cameroon	72	42	450			–	[108]
	Congo	7	–	1150.78	–	–	–	[109]
	Honduras	433	152	502	35	43	60	[110, 111]
	India	2631.93	21136.40	49382	–	4120.13	15.9	[97, 112, 113]
	Kenya	0.57	26	827	593	38	70	[95]
	Myanmar	50	420	3140	–	11640	7	[114]
	Pakistan	950	1250	7477		40	4	[115]
	Sudan	1000	500	1250	–	56	63.97	[116]
	Tunisia	25	245	62	–	–	3	[117]
	Ukraine	741.9	465.1	115	–	117.7		[118, 119]
	Vietnam	8	189.2	1648	–	270	4.4	[120]

Table 3.5 Renewable energy potential of different lower and lower-middle-income countries

Economic groups	Countries	Solar (MW)	Wind (MW)	Hydro (MW)	Geothermal (MW)	Bioenergy (MW)	References
Lower income countries	Ethiopia	3002	5200	45000	5000		[121]
	Liberia	–	–	2300	–	–	[122]
	Mali	1750		1150			[123]
	Rwanda	–	–	400	100	–	[96]
	Tanzania	800	50	3173	5000	–	[124]
	Uganda	–	–	2000	450	1650	[104]
Lower Middle-Income countries	Bangladesh	50174	20000	305	10000	2200	[125]
	Bolivia	–	108	40000	–	–	[126]
	Cameroon	500	80	1110	–	–	[108]
	Congo	85000		100000	–	–	[127, 128]
	India	200000	102000	20000	10000	23000	[113, 112, 129]
	Kenya	500	3000	3000	10000	1200	[130, 131]
	Myanmar	51793	–	39720	–	6000	[114]
	Pakistan	148677	43000	2958	100000		[132, 133]
	Sudan	1000	–	2927	400	–	[116, 134]
	Uzbekistan	1000		12000			[135, 136]
	Vietnam	7140	26763	7000	–	318630	[120]

times the current rate of production. Table 3.4 also indicates that geothermal energy is neglected in most of the nations; however, utilization of Solar and Hydro energy is much more common. Moreover, the utilization of wind and biomass energy needs much attention considering their significant potential.

Table 3.5 shows the renewable energy potential of different lower-income and lower-middle-income countries. From Table 3.5, it is clear that the nations of lower-income and lower-middle-income groups are not utilizing their full potential of renewable sources. Like Ethiopia has installed solar energy of only 14 MW while its potential for solar energy is 3002 MW, which indicates that the country has only utilized 0.47% of its total solar energy potential [98]. Hydro energy installed capacity of Ethiopia is 7.5 MW, though its potential of hydro energy is 5000 MW, which is inadequate compared to its potential. In the case of Tanzania, the installed capacity of hydro energy is 100 MW, while its potential for hydro energy is about 5000 MW. In percentage, the installed capacity is only 2% of the total hydro energy potential. For some countries like Liberia, Rwanda, and Sudan, the potential of different renewable sources is yet to be surveyed. This scenario of the utilization of renewable energy is insufficient. The government and policymakers of these countries should consider investing various renewable energy technologies to increase the percentage of renewable energy adoption.

From Table 3.5, it's clear that the lower and lower-middle-income countries, despite having a vast renewable energy potential, are unable to utilize it fully. Only renewable energy-based electricity production improves this situation. Efforts should be made to integrate renewable energy-based technologies in all sectors like household, commercial, industrial, transportation, etc. to use its true potential adequately. Utilizing the full potential of renewable energy will help to mitigate the future energy crisis of the lower and lower middle-income countries, as well as improve the living standards of its people. However, this book is especially focused on proposing renewable energy derived food preservation system (see Chap. 5 for details) in order to mediate the dependency on conventional energy sources for drying. Therefore, in the next section, an overview of the energy requirement for drying will be comprehensively represented. Furthermore, the challenges associated with renewable energy integration in the food drying sector will also be critically discussed.

3.3   Energy Required for Drying

Drying refers to the removal of water content from fresh food materials. Food drying is a proven technology that can effectively increase the shelf life of numerous perishable food products [137, 138]. However, maintaining an ideal condition for drying requires a substantial amount of energy. Therefore, even a diminutive improvement in energy efficiency in drying technologies may pave the way towards global energy sustainability. The energy requirement of different drying processes varies from each other. The energy requirement for drying depends on the type of technology applied and properties of food such as water content, structure, porosity, etc.

Among the major industrial processes, drying is undoubtedly the utmost energy-intensive technique [139]. Drying consumes almost 20–25% of the total energy used in the food processing industries [140]. Improving the energy efficiency of drying even by 1% could result in as much as a 10% increase in profit [139]. Moreover, there is a significant environmental concern, as energy for food drying comes mostly from conventional fossil fuel sources. The energy crisis is an impending threat to the sustainability of society. In order to produce adequate products and commodities for humans, the modern world relies heavily on the fossil fuel sources found in nature. Despite lacking regeneration capacity, the current infrastructure is draining the limited fossil fuel resources of the earth. It is evident from Fig. 3.6 that the most of the US energy comes from fossil fuel resources, which is around 80% of the total US energy if nuclear energy is considered as renewable energy sources [141]. Among many disadvantages of burning fossil fuel most egregious is the untold damage inflicted upon the eco-system through exorbitant carbon emission.

Almost all the countries of the world depend on fossil fuel for their energy demand, which is already discussed in this chapter. Figure 3.6 shows that the food processing industries of the United States of America (developed country) consume around 1.100 Quadrillion British thermal units (QBtu), and the majority of the US energy originates from fossil fuel resources [141]. The same scenario can be seen in the lower-income and lower-middle-income countries. As an example, in Fig. 3.7,

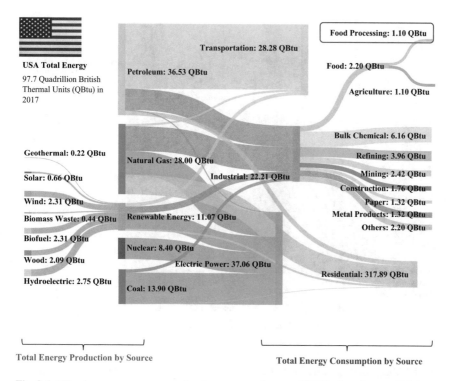

Fig. 3.6 US primary energy consumption by source and sector, 2017 Sankey diagram [141]

the consumption of energy in the food processing industries of Bangladesh (lower-middle-income country) is shown. Bangladesh consumes around 21.43 Trillion British thermal units (TBtu) for food processing, and most of its total usable energy comes from fossil fuel resources. This gives a clear view of how much energy is consumed by the food processing sectors of different countries irrespective of their position in development index. The scenario presented through these figures should create alarming concerns, as most countries are dependent on fossil fuel to satisfy their energy demand, which is plagued by problems like GHG emission and rapidly depleting reserve.

Figure 3.7 shows that in Bangladesh, only 9.85 TBtu comes from renewable sources, whereas the total energy generation is 1458.56 TBtu. Hence, it is evident that renewable energy is not fully utilized in this country. Therefore, renewable energy food processing technologies can potentially alleviate this gap.

The energy consumption in the food sector can be divided into various operations such as heating & drying, cooling, mechanical, infrastructure, and others. Figure 3.8 exhibits the percentage of energy usages in different processes of food industry.

From Fig. 3.8, it is clear that almost 60% of the total energy used in food industries is consumed by the heating and drying operation. Thus, it is essential to look for an alternative approach to intergrade renewable energy in the drying system of the food processing supply chain to reduce its demand for fossil-based energy.

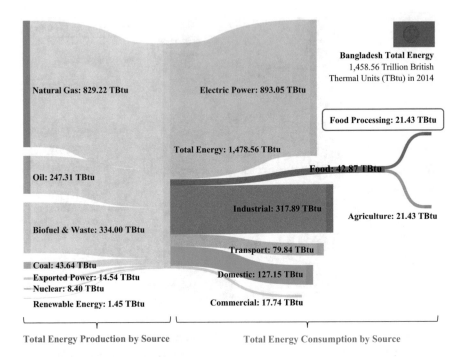

Fig. 3.7 Bangladesh primary energy consumption by source and sector, 2014 Sankey diagram [142]

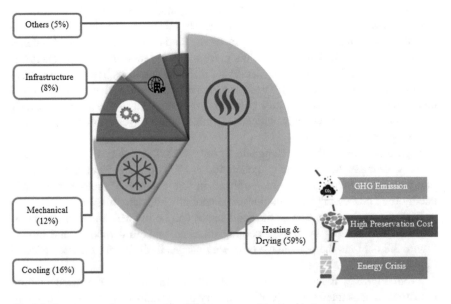

Fig. 3.8 Percentage of energy used in different sector of the food industry

Otherwise, due to excessive carbon emission and the greenhouse effect, earth's temperatures could reach an irreversible tipping point in just 12 years. The integration of renewable energy may face some critical obstacles that will make its mass utilization difficult using the current market available technologies. Barriers in the application of renewable energy are discussed in the following section.

3.4 Barriers in Renewable Energy Integration

Converting natural resources, known as renewable energy, into usable forms of energy should overcome some challenges. The underlying challenges of renewable energy technology integration in drying industries are summarized in Fig. 3.9. Among the challenges, technical and financial difficulties are the most important in the case of developing countries. If the cost of the competing alternatives (fossil fuel-based technologies) is lower than the cost of a proposed renewable energy-based drying techniques, then market penetration will be challenging.

All the other obstacles, including the technical barriers, societal barriers, infrastructural barriers, and marketing barrier, are known as non-economical barriers. Technology for enabling renewable energy integration is not fully developed, and more research & development is needed to make renewable energy integration easy and cheap. Therefore, technical barrier is critical to mass renewable energy utilization.

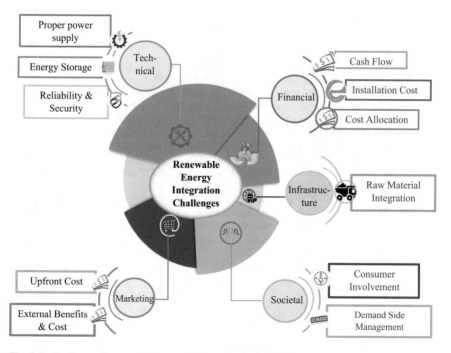

Fig. 3.9 Challenges in renewable energy integration in food drying

Incorporation of renewable energy in food drying technologies is not always possible in developing countries because challenges are more severe in developing countries compared to developed countries. More research & development can help to overcome the barriers and allow consumers to harness the full potential of renewable energy. If the challenges discussed in this section can be overcome, then it is possible to use the entire potential of renewable energy in the food drying sector. Comprehensive government policies, public-private investment and technological development may solve these challegens. Therefore, Chap. 5 of this book extensively discusses potential renewable energy-based sustainable drying technologies for developing countries. The techniques are chosen based on their financial and technical features, which complement the economy of developing countries. If developing countries integrate renewable energy-based drying technologies with their existing framework, it would significantly reduce the fossil fuel consumption related to drying.

References

1. M.U.H. Joardder, P.K. Halder, M.A. Rahim, M.H. Masud, Solar Pyrolysis: Converting Waste Into Asset Using Solar Energy, in *Clean Energy for Sustainable Development*, Elsevier, 213–235 (2017)
2. M.H. Masud, M.S. Akhter, S. Islam, A.M. Parvej, S. Mahmud, Design, construction and performance study of a solar assisted tri-cycle. Period. Polytech. Mech. Eng. **61**(3) (2017)
3. M.U.H. Joardder, R.J. Brown, C. Kumar, M.A. Karim, Effect of cell wall properties on porosity and shrinkage of dried apple. Int. J. Food Prop. **18**(10), 2327–2337 (2015)
4. M.U.H. Joardder, C. Kumar, M.A. Karim, Food structure: Its formation and relationships with other properties. Crit. Rev. Food Sci. Nutr. **57**(6), 1190–1205 (2017)
5. M.A. Karim, M. Hawlader, Development of solar air collectors for drying applications. Energy Convers. Manag. **45**(3), 329–344 (2004)
6. M.A. Karim, M.N.A. Hawlader, Performance evaluation of a v-groove solar air collector for drying applications. Appl. Therm. Eng. **26**(1), 121–130 (2006)
7. C. Sebi, "Explaining the increase in coal consumption worldwide," The Conversation (2019)
8. B. Dudley, "BP Statistical Review of World Energy 2019" (2019)
9. Worldcoal, "How Much Coal Is Left In The World , & When Will We Run Out ? How Much Coal Is Left In The World ?"(2019) Retrieved from Better Meets Reality website: https://www.bettermeetsreality.com/how-much-coal-is-left-in-the-world-when-will-we-run-out, Accessed date: June 4, 2019,
10. N. Abas, A. Kalair, N. Khan, Review of fossil fuels and future energy technologies. Futures **69**, 31–49, November (2015)
11. W. Islam, M. Mourshed, M.H. Masud, S.I. Sozal, S. Bin Sabur, Prospects of non-edible neem (Azadirachta indica) oil in Bangladesh: performance and emission evaluation in a direct injection diesel engine. Int. J. Ambient Energy **40**(5), 525–536 (2019)
12. A.R. Nabi, M.H. Masud, Q.M.I. Alam, Purification of TPO (Tire Pyrolytic Oil) and its use in diesel engine. IOSR J. Eng. **4**(3), 1 (2014)
13. IBRD, *The World Bank Data Catalog* (2018)
14. Robert Rapier, Asia's Insatiable Oil Demand, Forbes, 19th January (2018)
15. EIA, World oil statistics, (2016), Retrieved from: https://www.worldometers.info/oi, Accessed date: 5-06-2019
16. G. Kuo, When Fossil Fuels Run Out, What Then? (2019), Retrived from: https://mahb.stanford.edu/library-item/fossil-fuels-run, Accessed Date: 14-06-2019
17. EIA, Use of natural gas, (2019), Retrieved from: https://www.eia.gov/energyexplained/natural-gas/use-of-natural-gas.php Accessed Date: 22-06-2019

18. OECD, Global Gas Security Review 2019, 1st October (2019), Retrieved from: https://www.oecd.org/publications/global-gas-security-review-2019-9744bc3e-en.htm, Accessed date: 05-12-2019
19. EIA, Coal Reserves in the DR Congo, (2019), Retrieved from: https://www.worldometers.info/coal/democratic-republic-of-the-congo-coal/, Accessed Date: 22-06-2019
20. OPEC, Congo facts and figures, (2018), Retrieved from https://www.opec.org/opec_web/en/about_us/5090.htm, Accessed Date: 24-06-2019
21. W. Ahmed, Fossil fuel energy resources of Ethiopia, (2008)
22. Muluken Yewondwossen, Oil reserve in Ethiopia may hold up to 2 . 7 billion barrels, (2013), Retrieved from: https://www.awib.org.et/currency/news/item/322-oil-reserve-in-ethiopia-may-hold-up-to-2-7-billion-barrels.html, Accessed Date: 25-07-2019
23. Dawit Endeshaw, Ethiopia to Start Extracting Natural Gas, (2018), Retrieved from: https://allafrica.com/stories/201806270696.html, Accessed Date: 22-06-2019
24. Public Environmental Organization, Review of the Coal Sector in the Republic of Tajikistan, (2017).
25. U.S. Energy Information Administration, *Petroleum & Other Liquids* in Tajikistan, (2016), Retrieved from: https://www.eia.gov/international/data/country/TJK/infographic/petroleum-infographic?, Accessed Date: 29-06-2019
26. Pajhwok Afghan News, Oil in Afghanistan (2018)
27. Country Watch Incorporated, *Energy Consumption and Production Standard* (2012)
28. U.S. Embassies abroad, *Chad - Mining, Quarrying, and Oil and Gas Exploration* (2019), Retrieved from: https://www.privacyshield.gov/article?id=Chad-Mining-Quarrying-and-Oil-and-Gas-Exploration, Accessed date:19-07-2019
29. Nidhi Verma, N. F. *Yemen aims to export about 75,000 bpd oil in 2019: minister.*, Thomson REUTERS (2019)
30. CIA World Factbook, "Yemen Proved Reserves of Natural Gas," 2018, Retrieved from: https://www.indexmundi.com/yemen/natural_gas_proved_reserves.html, Accessed date: 20-07-2019
31. Engineer Mr. Salifou GADO, *The Energy Sector of Niger: Perspectives and Opportunities* (2015)
32. Index mundi, Niger Oil – proved reserves (2018), Retrieved from:https://www.indexmundi.com/niger/oil_proved_reserves.html, Accessed date: 20-07-2019
33. Economic Watch, *Benin Oil Production, Proven Oil Reserves, Consumption, Oil Exports and Oil Imports* (2019)
34. Index Mundi, Benin Natural gas - proved reserves, (2019), Retrieved from: https://www.indexmundi.com/benin/natural_gas_proved_reserves.html, Accessed Date: 17-02-2019
35. OPEC, "Angola facts and figures" (2019), Retrieved from:https://www.opec.org/opec_web/en/about_us/147.htm, Accessed date: 18-07-2019
36. Banglapedia, "Natural gas of Bangladesh" (2015), Retrieved from:http://en.banglapedia.org/index.php?title=Gas,_Natural, Accessed date: 21-07-2019
37. EIA, "Philippines Oil Production, Proven Oil Imports Reserves, Oil Consumption, Oil Exports and Oil," (2019), Retrieved from: https://www.worldometers.info/oil/philippines-oil/, Accessed Date: 22-07-2019
38. Enamul Hoque Chowdhury, Bangladesh holds largest oil, gas reserves in Asia Pacific (2016)
39. H. morgan, *Coal reserves - Country rankings,* (2019) Retrieved from: https://www.mining-technology.com/features/feature-the-worlds-biggest-coal-reserves-by-country/, Accessed Date: 17-09-2019
40. B. staton, Statista Accounts: Access All Statistics (2018)
41. Energía, "Bolivia estimates gas reserves will last beyond", (2017), Retrieved from: https://www.energia16.com/bolivia-estimates-gas-reserves-will-last-beyond-2032/?lang=en, Accessed Date: 15-07-2019
42. Ministry of coal, "Coal Reserves of India," (2018), Retrieved from: https://www.coal.nic.in/, Accessed Date: 19-07-2019
43. EIA, Reserves of Natural Gas in India, 2015, Retrieved from: https://www.worldometers.info/gas/india-natural-gas/, Accessed from: 20-07-2019

44. Global Economy, "Burma (Myanmar): Coal reserves" 2019, Retrieved from: https://www.theglobaleconomy.com/Burma-Myanmar/coal_reserves, Accessed date: 20-07-2019
45. Y. L. firens, "Burma (Myanmar): Oil reserves Last Chance to Invest at 500k", (2019)
46. Priv. Power Infrastruct. board, "Pakistan Coal Power Generation potential", (2004)
47. A. Hassan, "Pakistan' s Oil & Gas Reserves Will Run Out in 10 Years: Ministry", (2019).
48. F. P. Rahman, Pakistan Gas Reserve and Production, (2019), Retrieved from:http://www.gov.ph/ngas/ Accessed date: 22-07-2019
49. Priscilla Offiong, "Nigeria Relies on Oil Despite Having Large Coal Reserve", (2019), Retrieved from: https://www.climatescorecard.org/2019/05/nigeria-relies-on-oil-despite-having-large-coal-reserves/, Accessed date: 22-07-2019
50. OPEC, "Nigeria facts and figures," (2017), Retrieved from: https://www.opec.org/opec_web/en/about_us/167.htm, Accessed date: 22-07-2019
51. USAID Power Africa, "Democratic Republic of the Congo Energy Situation," (2019), Retrieved from: https://energypedia.info/wiki/Democratic_Republic_of_the_Congo_Energy_Situation, Accessed Date: 22-05-2019
52. IAEA, "Country Nuclear Power Profiles 2016 Edition" VIENNA (2016)
53. R. News, "Egyptian Crude Reserves at 2.4B Barrels", Egypt Oil & Gas Newspaper, 9th October (2017)
54. C. Stephen, "Egypt' s gas gold rush" Petroleum Economist, 28th February (2019)
55. Jeffrey Hays, Petroleum, Natural Gas and Coal in Vietnam, by Facts and details of Vietnam - Economic, Infrastructure and Energy, May (2014)
56. N. Van Thuyet, S.O. Ogunlana, P.K. Dey, Risk management in oil and gas construction projects in Vietnam. International journal of energy sector management , **1**(2), 175–194 (2007)
57. H. E Mohamed Arkab et. al., Algeria facts and figures, Annual Statistical Bulletin (2019), Retrieved from: http://www.opec.org/opec_web/en/about_us/146.htm., Accessed date: 14-05-2019
58. Algeria Data Portal, "Algeria Coal Reserves" International Energy Statistics, February (2015)
59. Chen Aizhu and Muyu Xu, "China opens up oil and gas exploration, production for foreign, domestic firms" Thomson REUTERS, 9th January (2020)
60. Chengcheng, China's coal hub discovers huge coal reserves, XINHUANET (2019)
61. EIA, Background Reference: Colombia, by US Energy Information Administration, 1–10 (2019)
62. CIA World Factbook, "Ecuador: Energy Resources", (2019), Retrieved from: https://openei.org/wiki/Ecuador, Accessed date: 22-07-2019
63. Organization of the Petroleum Exporting Countries, "Ecuador facts and figures" (2013)
64. EIA, "Guinea Coal", (2016), Retrieved from: https://www.worldometers.info/coal/guinea-coal, Accessed date: 23-07-2019
65. OPEC, "Equatorial Guinea facts and figures", (2018), Retrieved from: https://www.opec.org/opec_web/en/about_us/4319.htm, Accessed date: 14-06-2019
66. OPEC, "Gabon facts and figures", (2017), Retrieved from: https://www.opec.org/opec_web/en/about_us/3520.htm, Accessed date: 16-08-2019
67. OPEC, "Iraq facts and figures", (2019), Retrieved from: https://www.opec.org/opec_web/en/about_us/164.htm, Accessed date: 2-04-2019
68. OPEC, "Libya facts and figures", (2019), Retrieved from: https://www.opec.org/opec_web/en/about_us/166.htm, Accessed date: 11-07-2019
69. The GlobalEconomy, "Malaysia: Coal reserves", (2019) Retrieved from: https://www.theglobaleconomy.com/Malaysia/coal_reserves, Accessed date: 23-05-2019
70. International Energy Agency, "Malaysia Natural gas - proved reserves", (2018), Retrieved from: https://www.indexmundi.com/malaysia/natural_gas_proved_reserves.html, Accessed date: 24-07-2019
71. IndexMundi, "Malaysia Crude Oil Reserves by Year,", (2019), Retrieved from: https://www.indexmundi.com/energy/?country=my&product=oil&graph=reserves, Accessed date: 24-07-2019
72. NS Energy Staff Writer, "Countries with largest natural gas reserves in the Middle East", NS Energy (2019)

73. Jen Alic, "Turkey' s Oil Potential: Onshore and Offshore,", (2019)
74. Alex St John, Australian non-renewable energy resources, Parliament of Australia, (2019)
75. Australian Energy Statistics, "Australian Energy Update 2019", by Department of the Environment and Energy (2019)
76. World Coal Association, "Coal Facts", (2019), Retrieved from: https://www.nrcan.gc.ca/science-data/data-analysis/energy-data-analysis/energy-facts/coal-facts/20071, Accessed date: 24-07-2019
77. Natural Resources Canada, "Crude oil facts", (2012), Retrieved from: https://www.nrcan.gc.ca/science-data/data-analysis/energy-data-analysis/energy-facts/crude-oil-facts/20064, Accessed date: 24-06-2017
78. EIA, "Japan: Coal reserves", (2019), Retrieved from: https://www.worldometers.info/coal/japan-coal/, Accessed date: 24-07-2019
79. G. Leasequery, "Japan: Oil reserves", The Global Economy (2019)
80. CIA World Factbook, "Japan Natural gas - proved reserves", (2018), Retrieved from: https://www.indexmundi.com/japan/natural_gas_proved_reserves.html, Accessed date: 24-07-2019
81. OPEC, "Kuwait facts and figures", (2019), Retrieved from: https://www.opec.org/opec_web/en/about_us/165.htm, Accessed date: 24-07-2019
82. OPEC, "UAE facts and figures", (2019), Retrieved from: https://www.opec.org/opec_web/en/about_us/170.htm, Accessed date: 24-07-2019
83. Albilad, "Thar coal reserves is equivalent to total energy reservoirs of Saudi Arabia and Iran", (2018)
84. OPEC, "Saudi Arabia facts and figures", (2019), Retrieved from: https://www.opec.org/opec_web/en/about_us/169.htm, Accessed date: 24-07-2019
85. New Zealand Ministry of Economic Development, "New Zealand Oil Production and Reserves Statistics,", 2019, Retrieved from: https://openei.org/datasets/dataset/new-zealand-oil-production-and-reserves-statistics, Accessed date: 25-07-2019
86. Radio New Zealand, "NZ' s gas reserves drop, renewal unlikely due to ban on exploration," (2019)
87. The Oil & Gas Year, "Qatar: advantages in adversity" (2019), Retrieved from: https://theenergyyear.com/market/qatar/, Accessed date: 25-07-2019
88. EIA, "U.S. Coal reserves,", (2019)
89. Steven G. Grape, "U.S. Crude Oil and Natural Gas Proved Reserves, Year-end 2018, U.S. Energy Information Administration (2018)
90. Valerie Jones, "US Leads the World in Oil Reserves," RIGZONE, 12th June (2019)
91. Eurocoal, "The voice of coal in Europe: Spain", (2019), Retrieved from: https://euracoal.eu/info/country-profiles/spain/, Accessed date: 25-07-2019
92. Oil & Gas Authority, "UK Oil and Gas Reserves and Resources: as at end 2016," OGA Publication, (2017)
93. Centre for Climate and Energy Solution, Estimated Global Renewable Energy Share of Total Final Energy Consumption, (2017), Retrieved from: https://www.c2es.org/content/renewable-energy/, Accessed date: 12-09-2019
94. Trading Economics, "Uganda - Renewable electricity" 2019, Retrieved from: https://tradingeconomics.com/uganda/renewable-electricity-percent-in-total-electricity-output-wb-data.html, Accessed date: 25-07-2019
95. USAID, "Kenya' s power sector", Power Africa, (2018)
96. H. Eustache et al., Current Status of Renewable Energy Technologies for Electricity Generation in Rwanda and Their Estimated Potentials, Energy and Environmental Engineering, 6(1), 8–15 (2019)
97. Envecologic, "A Renewable Energy In India: Potential, Growth And Policies", (2017)
98. Mondal, M. A. H., Bryan, E., Ringler, C., Mekonnen, D., & Rosegrant, M., Ethiopian energy status and demand scenarios: Prospects to improve energy efficiency and mitigate GHG emissions, Energy, 149, 161–172 (2018)
99. M. Dorothal, "Ethiopia Solar Report The Solar Future: Deserts of Africa, Solar Plaza, 1–18, (2019)
100. Andrew Herscowitz, "Power Africa in Ethiopia, Ethiopia Energy Sector Overview" USAID, (2016)

101. Goanue, Augustus V. "Status of renewable energy in Liberia." Presentation of Rural and Renewable Energy Agency (2009)
102. Euromonitor International, "Mali: Country Profile," April (2019)
103. United Nations, Trade East African Community Facil, Tanzania country profile, (2013)
104. Kees Mokveld & Steven von Eije, "Final Energy report Kenya". Ministry of Foreign Affairs, (2018)
105. O.W.K. Avellino, F. Mwarania, A.H.A. Wahab, K.T. Aime, Uganda solar energy utilization: Current status and future trends. Published in international journal of scientific and research publications, **8**(3) (2018)
106. Md. Saidun Nabi, "Can Bangladesh meet its 10% renewable energy target by 2020?", Dhaka Tribune (2019)
107. M. Pansera, Renewable energy for rural areas of Bolivia. Renewable and sustainable energy reviews, **16**(9), 6694–6704 (2012)
108. L.L. Marius, N. Alex Joel, Energy sector of Cameroon. Africa Review, **11**(1), 34–45 (2019)
109. Energypedia, "Democratic Republic of the Congo Energy Situation Introduction", (2019), Retrieved from: https://energypedia.info/wiki/Democratic_Republic_of_the_Congo_Energy_Situation, Accessed date: 25-07-2019
110. E. C. Climático, "Solar energy: The revolution spurring development in Honduras" IDB, (2018)
111. H. C. Hayden, "A primer on renewable energy in Honduras" (2018), Retrived from: https://deetkenimpact.com/blog/renewable-energy-in-honduras/, Accessed date: 26-07-2019
112. S. Deepshikha, "Biomass Energy in India," Ministry of Foreign Affairs of Denmark (2016)
113. L. Tripathi, A.K. Mishra, A.K. Dubey, C.B. Tripathi, P. Baredar, Renewable energy: An overview on its contribution in current energy scenario of India. Renewable and Sustainable Energy Reviews, **60**, 226–233 (2016)
114. M.M. Tun, An overview of renewable energy sources and their energy potential for sustainable development in Myanmar. European Journal of Sustainable Development Research, **3**(1) (2019)
115. SEDA, "Renewable Energy in Malaysia" By Sustainable Energy Development Authority (SEDA) Malaysia (2019)
116. A. M. Omer, E. Z. Braima, Renewable Energy Potential in Sudan. International Energy Journal, 2(1) (2007)
117. Energypedia, "Tunisia Energy Situation", (2017), Retrieved from: https://energypedia.info/wiki/Tunisia_Energy_Situation, Accessed date: 22-07-2019
118. Mr. Maksim Urakin, "Ukraine Increases Electricity Generation from Renewable Energy Sources by 2.8 Times" (2019), Retrieved from: https://open4business.com.ua/ukraine-increases-electricity-generation-from-renewable-energy-sources-by-2-8-times/, Accessed date: 12-07-2019
119. S. Teush, Ukraine' s renewable energy outlook under the new electricity market design, pv magazine, 1–12, (2018)
120. Koushan Das, "Renewables in Vietnam: Current Opportunities and Future Outlook", (2018), Retrieved from: https://www.vietnam-briefing.com/news/vietnams-push-for-renewable-energy.html/, Accessed date: 25-07-2019
121. Maasho, A. Ethiopia opens Africa's largest wind farm to boost power production'. Thompson Reuters, 26. 1–9 (2013)
122. Energypedia, Liberia Energy Situation, (2016), Retrieved from: https://energypedia.info/wiki/Liberia_Energy_Situation, Accessed date: 25-07-2019
123. Energypidia, Mali Energy Situation. 1–8, (2019), Retrieved from: https://energypedia.info/wiki/Mali_Energy_Situation, Accessed Date: 12-06-2019
124. O.K. Bishoge, L. Zhang, W.G. Mushi, The potential renewable energy for sustainable development in Tanzania: A review. Clean Technologies, **1**(1), 70–88 (2019)
125. P.K. Halder, N. Paul, M.U. Joardder, M. Sarker, Energy scarcity and potential of renewable energy in Bangladesh. Renewable and Sustainable Energy Reviews, **51**, 1636–1649 (2015)

126. State of Green, Bolivia set to quadruple its energy production from wind power, Ministry of Foreign Affairs Denmark, The Danish Investment Fund for Developing Countries (2018), Retrieved from: https://stateofgreen.com/en/partners/state-of-green/news/bolivia-set-to-quadruple-its-energy-production-from-wind-power/, Accessed date:26-07-2019

127. R. Deshmukh et al., Renewable Riches: How Wind and Solar Could Power DRC and South Africa, International Rivers, September (2017)

128. USAID, Democratic Republic of the Congo, (2013), Retrieved from: https://www.usaid.gov/powerafrica/democratic-republic-congo, Accessed date: 26-07-2019

129. N.K. Sharma, P.K. Tiwari, Y.R. Sood, Solar energy in India: Strategies, policies, perspectives and future potential. Renewable and Sustainable Energy Reviews, **16**(1), 933–941 (2012)

130. Pacifica Ogola and Luis Patron, Making a Difference Through Geothermal Energy, United Nations University, 12th September (2019)

131. MOE, Renewable Energy potential map by Ministry of the Environment, 1–10, (2018), Retrieved from: https://www.env.go.jp/earth/ondanka/rep/EN/index_en.html, Accessed Date; 30-05-2019

132. M.K. Farooq, S. Kumar, An assessment of renewable energy potential for electricity generation in Pakistan. Renewable and Sustainable Energy Reviews, **20**, 240–254 (2013)

133. M. Shakir, I.U. Haq, M.A. Khan, S.A. Malik, S.A. Khan, Alternate energy resources for Pakistan: sustainable solutions for fulfilling energy requirements. World Applied Sciences Journal, **31**(5), 718–723 (2014)

134. J. Deng, Regional motivation to develop South Sudan's hydro power capacity. Africa's Power J., 1–8 (2019)

135. E.Y. Rakhimov, S.E. Sadullaeva, Y.G. Kolomiets, K.K. Tashmatov, N.O. Usmonov, Analysis of the solar energy potential of the Republic of Uzbekistan. Applied Solar Energy, **53**(4), 344–346 (2017)

136. Eurasianet, *Uzbekistan plans its own hydropower plant* (2017), Retrieved from:https://eurasianet.org/uzbekistan-plans-its-own-hydropower-plant, Accessed date: 27-07-2019

137. M.U. Joardder, C. Kumar, M.A. Karim, Multiphase transfer model for intermittent microwave-convective drying of food: Considering shrinkage and pore evolution. International Journal of Multiphase Flow, **95**, 101–119 (2017)

138. C. Kumar, M.U.H. Joardder, T.W. Farrell, G.J. Millar, M.A. Karim, Mathematical model for intermittent microwave convective drying of food materials. Drying Technology, **34**(8), 962–973 (2016)

139. T. Kudra, Energy aspects in drying, Drying Technology, 22(5), 917–932 (2004)

140. K.J. Chua, A.S. Mujumdar, M.N.A. Hawlader, S.K. Chou, J.C. Ho, Batch drying of banana pieces—effect of stepwise change in drying air temperature on drying kinetics and product colour. Food Research International, **34**(8), 721–731 (2001)

141. U.S. Energy Information Administration, "Use of Energy in the United States Explained," (2019)

142. S. Islam, M.Z.R. Khan, A review of energy sector of Bangladesh. Energy Procedia, **110**, 611–618 (2017)

Chapter 4
Practiced Drying Technologies in Developing Countries

Drying is one of the most extensively used forms of technology for the preservation of food. Conventional food preservation techniques such as freezing require an immense amount of energy that is sometimes unmanageable, particularly for the low-income developing countries of the world. Due to the limited economic capability, they rely on inexpensive conventional food preservation techniques.

Though existing food drying technologies can range to various prices, the populace of developing countries generally is accustomed to using inexpensive options due to limited resource availability and expenditure capacity. Though expensive drying technology provides better drying speed and excellent dried food quality, they are not always accessible in developing countries for financial reasons. As such, drying technologies that are cheap, easy to operate, and effective are more common in developing countries. The conventional drying technologies meeting the requisites of the general people of these countries, especially in terms of economic viability, will be discussed in this chapter.

Sun-drying is the most common and familiar form of drying technique in developing countries as it needs a minimum investment to apply it [1, 2]. Although the dried food quality is not satisfactory due to its unhygienic approach, people of developing countries mostly depend on this technology due to their financial limitations. This indicates that economic viability plays an important role when selecting a drying technology in developing countries, such as Bangladesh. Therefore, when proposing a new drying technique for them, one should keep in mind that the technology needs to be economically viable for its customers. Furthermore, power, fuel, and raw material scarcity also have an undeniable influence on the type of drying technologies used in developing countries. Figure 4.1 describes the key features that are required to develop a food dryer for the low-income developing countries of the world. The conventional drying technologies are generally fabricated using commonly available raw materials like polymer plastic, glass, straw, etc. in order to reduce the manufacturing cost of the dryer. For example, the drying chamber of an STR dryer is made of straw. When developing

© Springer Nature Switzerland AG 2020
M. Hasan Masud et al., *Sustainable Food Drying Techniques in Developing Countries: Prospects and Challenges,*
https://doi.org/10.1007/978-3-030-42476-3_4

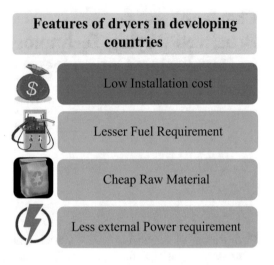

countries adopt drying technology, it is crucial to meet the requirements outlined in
Fig. 4.1. The researchers are putting in rigorous efforts to develop new drying
technologies that can meet up the demand of developing countries. However, there
are many opportunities to develop new energy-efficient dryers for developing
countries.

Generally, five types of dryers are most commonly used in developing countries.
Figure 4.2 breaks down their crucial features, which are partially fulfilling the needs
of developing countries. However, research on creating more advanced energy-
efficient dryers is essential to provide a better life for the people of the developing
countries.

In the following subsections, the underlying mechanisms of conventional drying
technologies are discussed in great detail. In addition, the advantages and
disadvantages of these technologies are also presented to identify further
improvements that can be made. This will allow the consumer to select the
appropriate dryer for their specific use.

4.1 Sun Drying

Sun drying is an ancient and widespread form of drying that is practiced in most
developing countries of the world because of its simple design and easy operation
[3]. The drying of corn by the process of sun-drying in a rural area of Bangladesh is
shown in Fig. 4.3.

Sun-drying does not need much technological support, as only keeping food in
the sunlight is sufficient to dry them. Sun-drying techniques can be classified in
different ways, as shown in Fig. 4.4.

Fig. 4.2 Dryers used in developing countries in general and their features

4.1.1 Basic Principle of Sun Dryer

In the process of sun drying, thermal energy from the sun is used. The heat from the sun evaporates the water particles in the outer layer of the food. As the water content of the external layer decrease, the water from inner layers of the food move to the outer layer through osmosis, and the evaporation process continues. The surrounding airflow replaces the wet air with dry air to increase the rate of evaporation. Figure 4.5 displays the sun drying process.

Fig. 4.3 Sun drying of corn (maize)

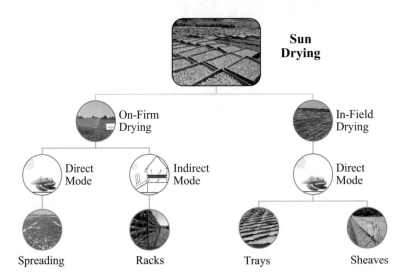

Fig. 4.4 Sun-drying classification

Although sun-drying is a widely used technique, it has some drawbacks as well. The following disadvantages make the system inconvenient and show that the development of improved solar drying techniques is crucial [4, 5]:

- Dust and microorganism in the air makes the process unhygienic
- Sun-drying cannot be done in cloudy weather and during rain
- Sun-drying needs continuous monitoring to protect food from wild birds and animals

Fig. 4.5 Basic working process of sun-drying technology

- The dried food quality is deteriorated due to microorganism and insects
- The demand for the land for sun-drying is more than three times bigger compared to solar drying
- In the case of direct sun drying, Ultraviolet (UV) radiation can damage the quality of food

Despite these drawbacks, sun drying is still widely used in many developing countries, because its inexpensive and straightforward method outweigh the disadvantages.

4.2   Solar Drying

Solar drying is an improved version of sun drying. Both drying techniques use solar energy; however, solar drying does not accomplish its task in the open air [6]. Longer shelf life and faster rate of drying and protection (from rain, insects, and bacteria) make solar drying more popular in contrast to open-air sun drying [7]. Solar drying can be classified in the following categories, as represented in Fig. 4.6 [8]. Figure 4.7 shows a solar dryer in operation.

4.2.1   Typical Features of Solar Dryers

Among the different categories of solar dryers, four types are most commonly used in developing countries. The typical features of these four types are described below:

Active Solar Dryer In these solar dryers, airflow is conducted in a forced method. A fan or blower is used to circulate the airflow through the solar collector and food chamber [8].

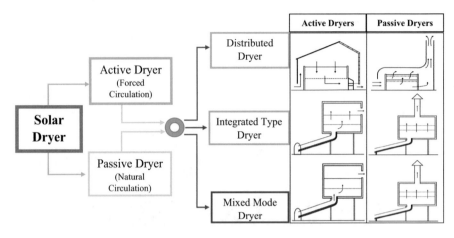

Fig. 4.6 Classification of solar drying techniques

Fig. 4.7 Example of a solar dryer

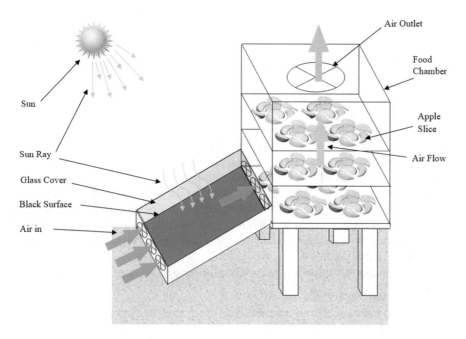

Fig. 4.8 Working principles of an indirect solar dryer

Passive Solar Dryer In a passive solar dryer, the air is heated and circulated naturally by buoyancy force, wind pressure, or a combination of both. No fan or blower is used to facilitate the airflow [8].

Mixed-Mode Solar Dryer In these dryers, both sun rays and heated air from solar collectors are used to drive the drying process. While the sunrays directly impose on the food, the hot air from the solar collector blows through the food to increase the drying rate [9].

Hybrid Solar Dryer This technology is similar to the mixed-mode technique. However, in this type of dryer, the fan which maintains the airflow is run by the solar power PV cell [9].

When choosing a dryer or drying method, the type, availability, cost, energy consumption, and the final quality of the dried product need to be taken into account [3]. However, for developing countries, the choice is fundamentally influenced by the requirement of energy, and as said before, the cost. The basic principles of solar drying, along with the underlying mechanism, is described in Fig. 4.8.

4.2.2   Basic Principles of Solar Dryer

The solar drying technique is carried out by collecting solar energy through heating the air volume in a confined solar collector. This heated air from the solar collector is blown through the drying chamber. The food is kept in the chamber, and drying takes place by convective heat transfer and evaporative mass transfer. The black surface in the dryer absorbs the solar radiation. The black color is perfect for absorbing the energy from solar radiation. The solar collector is covered by glass, which helps to create a greenhouse effect, i.e., it permits the sunray to come in but impedes the reflected sun ray to leave the collector. As a result, the air temperature inside the solar collector increases. The air enters the collector by the air inlet and subsequently heats up by the convection mechanism. The heated air is then passed to the drying chamber where the food is placed. After passing heat to the food, the air is released to the atmosphere by an air outlet in the top surface of the food chamber, as shown in Fig. 4.8.

There are also many drawbacks in the solar drying technique, including [10, 11]:

- Full dependency on solar energy
- Dust and microorganisms can enter and reduce the quality of dried food
- High cost compared to sun drying
- Needing maintenance after a particular period
- Necessitates hot and dry climate (preferable ambient relative humidity is 60%)
- Sophisticated and complex technology compared to sun drying

Though solar drying has some drawbacks, the advantages of solar drying techniques prevail over the limitations, and, therefore, this technology is extensively used in developing countries. The simplicity, cheap investment, and easy operation of this technology, along with its renewable application, make solar drying popular in developing countries.

All of the solar drying systems have their unique advantages and specific strength over others. Some dryers can function longer, whereas some others can process a bulk quantity of agricultural products. In order to select a proper solar drying technique, it is imperative to delineate the effectiveness of each dryer. Based on the performance analysis, an effectiveness heat map has been constructed, which is revealed in Table 4.1. It is expected that the developed map will help the consumers to select the appropriate dryer for their use.

4.3   Biomass Drying

According to the International Union of Pure and Applied Chemistry (IUPAC), biomass is a material that is produced by the growth of microorganisms, plants, or animals, which is considered as one of the significant sources of renewable energy. Agriculture crop residue, forest residue, animal manure, municipal solid waste, and

Table 4.1 Heat map of different solar dryers depending on drying speed, time, dried food quality, cost, and environmental effect

Solar Dryer Type	Drying Speed	Drying Time	Dried Food Quality	Cost	Environmental effect
Active Solar Dryer	Low	High	Low	Low	Low
Passive Solar Dryer	Medium	Medium	Medium	Medium	Medium
Mixed-mode Solar Dryer	High	Low	High	High	Medium
Hybrid Solar Dryer	High	Low	High	Medium	Low

Table 4.2 Research in biomass drying improvement

Source	Year	Location	Application	References
B. Bena et al.	2002	Australia	Vegetables	[18]
A. Madhlopa et al.	2007	Malawi	Pineapple	[19]
K. Gunasekaran	2012	India	Coleus Forskolin stems	[20]
S. Dhanushkodi et al.	2015	India	Cashew	[21]

sewage are widely available forms of biomass [12–14]. There is a vast potential of biomass present in the developing countries of the world. A comprehensive study on municipal solid waste and animal manure shows the substantial potential of biomass in the developing countries of the world, which can be promptly used for food drying [15–17]. A significant amount of heat energy is released by burning the biomass materials, which can subsequently be used for the drying of food materials by the principle of biomass drying. Biomass dryers are being widely used in developing countries for its easily accessible feed materials and uncomplicated characteristics.

From the invention of biomass drying technology, it has gone through several stages of improvement. Researchers have tried to improve its effectivity and efficiency by incorporating different ideas, and today's improved biomass dryer is a result of those inclusive researches and developments. Some works which have a contribution to the improvement of this drying technology is summarized in Table 4.2. Biomass dryer being very popular with the developing countries, it is also used at the industrial level. BAIXIN company has developed one of the most efficient biomass food dryer (Model: 3.5 t/h HGX-7).

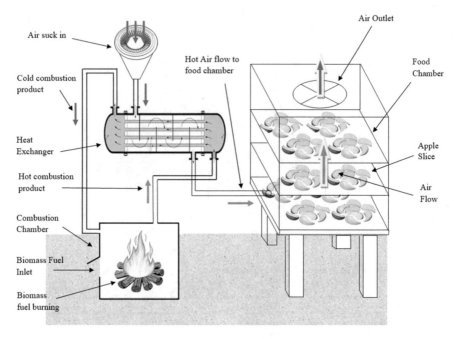

Fig. 4.9 Biomass food drying technique

4.3.1 Basic Principle of Biomass Dryer

In biomass dryers, the energy from the combustion of biomass is used as a source of heat for drying the food materials. The biomass dryers consist of a burner where biomass materials such as lumber, debris, manure, etc. are burnt. In the case of direct biomass dryer, the product of combustion is directly flown through the food chamber, where the food products to be dried are kept. The food is dried in the process of convection heat transfer. Based on the principle of evaporation mass transfer, the moisture is removed from the outer layer of the food products. As a result, the moisture content of the outer layer decreases. As such, the water of the inner layer of the food flows to the outer layer by osmosis through the cell membrane. Then the evaporation mass transfer of this water takes place. All the above processes take place continuously until the food dries to the desired level.

In the case of indirect biomass food dryer, the heat of the combustion product is used to heat the secondary working fluid, usually fresh air. This is usually done by using a heat exchanger. This working fluid is then flown inside the food chamber where the food is placed. The process of indirect biomass drying is presented in Fig. 4.9. The indirect drying process is more preferable and hygienic as the harmful content of combustion products such as Carbon Dioxide, Carbon Monoxide, Sulfur Dioxide, Nitrogen Oxides, Lead, Particulate Matter, etc. do not flow inside the food chamber and contaminate the food products.

One major problem faced by biomass food dryers is the burning of outer cells of food products. The flow of moisture from the inner layer to the outer layer of the food by osmosis takes a significant amount of time. As a result, the food products do not get enough time for the water to flow from the inner layer to the outer layer, and the burning of the outer layer cells of the food products occurs. As it is not convenient to extinguish the fire in the burner and ignite it again continuously to flow the heated air when necessary, this problem is a matter of concern.

Biomass food drying has some disadvantages alongside its implausible advantages. These disadvantages highlight the necessity of proposing improved biomass technology. The problems associated with biomass food drying are [19]:

- CO_2 emission
- High capital cost
- Toxicity of combustion product diminish dryer food quality
- Requires periodical maintenance
- Sophisticated and complex technology

Despite having the above disadvantages, a biomass food dryer is still a popular form of drying in developing countries because of the availability and inexpensiveness of feed materials used in the dryer to produce heat energy for drying.

4.4   STR Dryer

STR dryer is a low-cost batch dryer, which was developed by Centre for Agricultural Energy and Machinery, Nong Lam University, Vietnam, and Japan International Research Center for Agricultural Sciences (JIRCAS). In lower-middle income developing countries like Bangladesh, STR is generally used for drying rice [22]. Figure 4.10 shows an STR dryer used in Bangladesh Agricultural University (BAU).

In accordance with the trend of improving product quality and efficiency, many researchers have studied various features of STR drying. Results from their experiment have improved the utility of STR drying technologies in developing countries. Major researches on the performance of STR drying are summarized in Table 4.3.

4.4.1   Basic Principle of STR Dryer

The STR dryer consists of two perforated (inner and outer) bins having annular space between them, where food to be dried is placed. The inner bin diameter is fixed, while the outer bin diameter is adjustable depending on the volume of food. The top and bottom of the annular space are closed. Biomass stoves or other kinds of stoves are generally used as a source of heat. An axial flow blower is used to draw the hot air from the stove through a pipe and force it radially through the perforated bins, as shown in Fig. 4.11. The food is dried in the annular space by this hot air. The

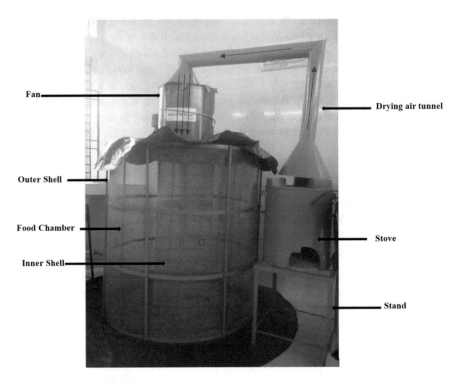

Fig. 4.10 STR dryer in use

Table 4.3 Overview of research on STR drying

Source	Year	Location	Application	References
MA Alam et al.	2015	Bangladesh	Grain bin	[23]
Ashok Kumar et al.	2017	India	Paddy	[24]
R. Bhadra et al.	2017	USA	Grain	[25]
CK Saha et al.	2017	Bangladesh	Paddy	[26]
MA Alam et al.	2018	Bangladesh	Rice	[27]
MA Alam et al.	2019	Bangladesh	Paddy	[28]

hot air from stove provides heat to the food by convection process, and moisture removal is occurred by evaporation process. This type of dryer needs two categories of energy; one is for fuel that will burn in the stove, and the other is electricity to run the axial flow blower.

The direct air from the stove contains harmful contaminants with it, which can get mixed with the food and deteriorate its quality. Therefore, the direct air pass from the stove should be replaced with a heat exchanger, which can significantly improve the dried food quality. However, the heat exchanger has the problem of ineffectively using heat energy owing to its improper design and consequently increases the cost of drying.

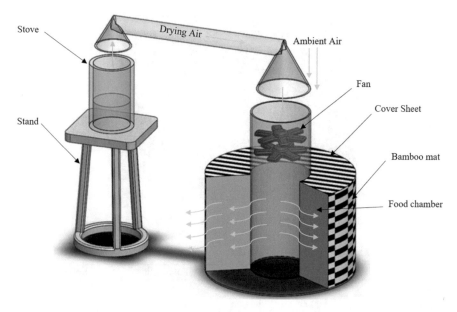

Fig. 4.11 STR dryer working principle

Disadvantages which threaten its usability in the developing countries and pave the path for finding new drying technologies are listed below [24, 28]:

- The fuel used is expensive compared with other conventional low-cost drying technologies
- Continuous electricity supply is necessary to run the blower
- High capital cost is required
- Controlling of heat source is difficult
- The CO_2 emission from the stove
- The heated air might contain microorganism and dust which may deteriorate the dried food quality

Although the STR dryer has the above drawbacks, it is one of the most common drying technologies that is used in developing countries for grain drying. Reasons why this technology is still prevalent in developing countries include simplicity of the dryer, high effectivity and capacity, and reasonable operating cost.

4.5 Reversible Air Dryer, SRA

The SRA is a Vietnamese reversible dryer series. Here, RA is an abbreviation for Reversible-Air, S is drying in Vietnamese. The unique feature which makes this dryer different from others is that the dying is reversed only once, and less reversal means less involvement of manual labor. Twenty-five SRA units have been applied

Air Duct

Generator to
run axial
flow fan

Grain

Food Chamber

Fig. 4.12 The SRA-1.5 reversible air dryer in practical operation in Vietnam

Table 4.4 Research on SRA dryer development

Source	Year	Location	Application	References
P Kuizon et al.	1996	Philippines	Rice	[32]
PH Hein et al.	2000	Vietnam	Grain	[33]
Htam Nguyen et al.	2002	Vietnam	–	[34]
PH Hein et al.	2003	Vietnam	Paddy & coffee	[31]
PH Hein et al.	2003	Vietnam	Grain	[35]
CJ Tado et al.	2015	Philippines	Grain	[36]

successfully in various provinces of Vietnam [29]. Developing countries like
Taiwan, Philippines, and Vietnam are successfully using reversible air dryers on a
large scale [30, 31]. The practical operation of SRA-1.5 reversible air dryer in
Vietnam is represented in Fig. 4.12.

In recent times, significant research has been done on the development of the
SRA dryer. Some inclusive studies, which have contributed to the progress of SRA
dryer development, are summarized in Table 4.4.

4.5.1 Basic Principle of SRA Dryer

SRA consists of a reversible side-duct plenum chamber, capable of reversing the
drying air. It also provides a low drying bed, which is convenient for loading and
unloading. A two-stage axial flow fan is used to provide reversible airflow through
the duct in the drying chamber. In the drying chamber, a perforated drying bed is
used in order to dry the food. Figure 4.13 is the pictorial representation of the
working principle of SRA (reversible air) dryer. Drying air is flown to the drying
chamber by the two-stage axial flow fan. The airflow is reversed using one-way
valves, as shown in Fig. 4.13. The bed in the drying chamber has paths for air to

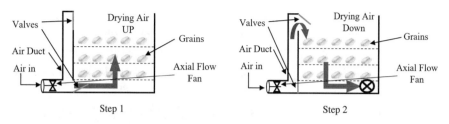

Fig. 4.13 SRA reversible air dryer operating principle

flow through the food products. The reversion of the airflow is done with a continuous interval. As a result of constant inversion, the drying air flows to the food products from all directions, and evaporation from all the surfaces of the food takes place. Therefore, the rate of drying increases significantly.

Though having numerous advantages to be an ideal drying technology for developing countries, the SRA also has some demerits. That is why new dryer technologies for developing countries are essential. The drawbacks of SRA reversible air dryer include [35, 36]:

- High initial cost
- The effectiveness of dryer is dependent on initial moisture content
- Need periodic maintenance
- Continuous electricity supply is necessary to run the fan
- It can only be performed while the humidity of the air is low

However, besides its disadvantages, the outstanding features which made this dryer suitable for developing countries are its simplicity, high capacity, and low operating cost. For these reasons, SRA reversible air dryer is being used in developing countries.

The effectiveness of the above-described drying technologies varies depending on the types of food dried. Some drying technology suit only one food group while others suit more than one food group. For instance, sun-drying supports drying of all the four groups of food, i.e., cellular, fibrous, crystalline, and gel, while both solar and biomass drying best suit for cellular, fibrous and crystalline food drying [37]. On the other hand STR and SRA drying technologies are only suitable for fibrous food like paddy [26] [29].

The above-discussed dryers are popular because their features meet the needs of developing countries. Some techniques are cheap but not much efficient, while some are effective but costly. So, taking these into consideration, people use the technology which suits their purpose and financial capability. Therefore, there is a need for a comparative study of these existing drying technologies. Table 4.5 represents a heat map that is prepared for a comparative study of the existing drying technologies depending on cost, energy requirement, dried food quality, drying speed, and environmental effect. It is important to note that the scale of delineation is limited to only among the five existing drying techniques discussed in Chap. 5.

Table 4.5 Heat Map of different types of dryer used in developing countries based on cost, required energy, dried food quality, drying speed and environmental pollution

Drying Technology	Cost	Energy Required	Dried Food Quality	Drying Speed	Environmental Pollution
Sun Drying	Low	Low	Low	Low	Low
Solar Drying	Medium	Low to Medium	Medium	Medium	Low
Biomass Drying	Medium	Medium	Low to Medium	High	High
STR Dryer	High	High	High	High	High
Reversible air dryer SRA	Medium to High	Medium	Medium to High	High	Low

This chapter is focused on investigating the current drying scenario of developing countries. Thus, comparing the performance of industrial or advanced drying technologies is outside the scope of this study. Therefore, the impact scale, which ranges from low to high is specifically defined to compare the performance among only the following five drying techniques of developing countries. The drying techniques considered are sun drying, solar drying, biomass drying, STR dryer, and SRA dryer.

These existing drying technologies are so far appeasing the need of developing countries. But with the development of technology and steady growth in population, the requirement of new and efficient drying technology is increasing substantially. The increased research and development have helped people to know about the insights of drying more precisely. Currently, there is a demand for speedier drying with low running costs and higher dried food quality. Consequently, people are trying to find environment friendly technology while being efficient and faster. The impending crisis of conventional fuel sources has added a new dimension to this requirement. The people of developing countries have become more conscious about their health. Therefore, the dried food quality has become a critical issue among them. All these requirements are acting as a driving force for finding new, efficient, effective, faster, and environment friendly drying technology for developing countries. Therefore, Chap. 5 of the book will propose some energy-efficient improved drying technologies for developing countries with the potential to operate on renewable energy sources.

References

1. M.A. Karim, M.N.A. Hawlader, Drying characteristics of banana: Theoretical modelling and experimental validation. J. Food Eng. **70**(1), 35–45 (2005)
2. M.A. Karim, M. Hawlader, Development of solar air collectors for drying applications. Energy Convers. Manag. **45**(3), 329–344 (2004)
3. V.R. Sagar, P.S. Kumar, Recent advances in drying and dehydration of fruits and vegetables: A review. J. Food Sci. Technol. **47**(1), 15–26 (2010)
4. M.A. Karim, M.N.A. Hawlader, Performance evaluation of a v-groove solar air collector for drying applications. Appl. Therm. Eng. **26**(1), 121–130 (2006)
5. M.A. Karim, M.N.A. Hawlader, Performance investigation of flat plate, v-corrugated and finned air collectors. Energy **31**(4), 452–470 (2006)
6. F. Senatore, F. Napolitano, M.A. Mohamed, P.J.C. Harris, P.N.S. Mnkeni, J. Henderson, Antibacterial activity of Tagetes minuta L.(Asteraceae) essential oil with different chemical composition. Flavour Fragr. J. **19**(6), 574–578 (2004)
7. S. Janjai et al., Experimental and simulated performance of a PV-ventilated solar greenhouse dryer for drying of peeled longan and banana. Sol. Energy **83**(9), 1550–1565 (2009)
8. A. Balasuadhakar, A review on passive solar dryers for agricultural products. İnt. J. Innov. Res. Sci. Technol. **3**(01), 64–70 (2016)
9. M.A. Khan, M.S. Sabir, M. Iqbal, Development and performance evaluation of forced convection potato solar dryer. Pakistan J. Agric. Sci. **48**(4), 315–320 (2011)
10. M.G. Green, D. Schwarz, Solar drying technology for food preservation. Energy **49**(0), 1–8 (2001)
11. A. Shakerardekani, R. Karim, H.M. Ghazali, N.L. Chin, Types of dryers and their effect on the pistachio nuts quality-a review. J. Agric. Sci. **3**(4) (2011)
12. A.S.N. Huda, S. Mekhilef, A. Ahsan, Biomass energy in Bangladesh: Current status and prospects. Renew. Sust. Energ. Rev. **30**, 504–517 (2014)
13. N. Newaj, M.H. Masud, Utilization of waste plastic to save the environment. In International conference on mechanical, industrial and energy engineering, KUET, Khulna, Bangladesh, 1-4 (2014)
14. M.H. Masud, R. Ahamed, M.U.H. Joardder, M. Hasan, Mathematical model of heat transfer and feasibility test of improved cooking stoves in Bangladesh. Int. J. Ambient Energy **40**(3), 317-328 (2019)
15. R.J. Haynes, R. Naidu, Influence of lime, fertilizer and manure applications on soil organic matter content and soil physical conditions: A review. Nutr. Cycl. agroecosystems **51**(2), 123–137 (1998)
16. M. Mourshed, M.H. Masud, F. Rashid, M.U.H. Joardder, Towards the effective plastic waste management in Bangladesh: a review. Environmental Science and Pollution Research, **24**(35), 27021-27046 (2017)
17. M.H. Masud et al., Towards the effective E-waste management in Bangladesh: A review. Environ. Sci. Pollut. Res. **26**(2), 1250–1276 (2019)
18. B. Bena, R.J. Fuller, Natural convection solar dryer with biomass back-up heater. Sol. Energy **72**(1), 75–83 (2002)
19. A. Madhlopa, G. Ngwalo, Solar dryer with thermal storage and biomass-backup heater. Sol. Energy **81**(4), 449–462 (2007)
20. K. Gunasekaran, V. Shanmugam, P. Suresh, Modeling and analytical experimental study of hybrid solar dryer integrated with biomass dryer for drying Coleus Forskohlii stems, in *2012 IACSIT Coimbatore Conferences*, **28**(1), 28–32 (2012)
21. S. Dhanushkodi, V.H. Wilson, K. Sudhakar, Design and performance evaluation of biomass dryer for cashewnut processing. Adv. Appl. Sci. Res. **6**, 101–111 (2015)
22. M.A. Alam, C.K. Saha, M.M. Alam, M.A. Ashraf, B.K. Bala, J. Harvey, Neural network modeling of drying of rice in BAU-STR dryer. Heat Mass Transf. und Stoffuebertragung **54**(11), 3297–3305 (2018)

23. M.A. Alam, C.K. Saha, M.A. Momin, M.M. Alam, B.K. Bala, Spatial distribution of temperature and moisture in grain bin and grain bin size effect on STR dryer performance in Bangladesh. J. Agril. Mach. Bioresour. Engg 7(1), 1–8 (2016)
24. A. Kumar, S. Kumar, S. Kumar, P. Kalita, K. Rausch, Effect of hot air velocity on drying kinetics undergoing str drying of paddy. Int. J. Sci. Environ. Technol. 6(5), 2735–2743 (2017)
25. R. Bhadra, Reducing post-harvest loss in developing countries through the feed the future initiative. Resour. Mag. 24(3), 12–15 (2017)
26. C.K. Saha, M.A. Alam, M.M. Alam, P.K. Kalita, J. Harvey, Field performance of BAU-STR paddy dryer in Bangladesh, in *2017 ASABE Annual International Meeting*, 1 (2017)
27. M.A. Alam, C.K. Saha, M.M. Alam, M.A. Ashraf, B.K. Bala, J. Harvey, Neural network modeling of drying of rice in BAU-STR dryer. Heat Mass Transf. 54(11), 3297–3305 (2018)
28. M.A. Alam, C.K. Saha, M.M. Alam, Mechanical drying of paddy using BAU-STR dryer for reducing drying losses in Bangladesh. Progress. Agric. 30, 42–50 (2019)
29. H.H. Phan, H.T. Nguyen, N. Van Xuan, Study on the reversal timing for the SRA reversible dryer, *Agric. Eng. agroproducts Process. Towar. Mech. Mod. Rural areas*, no. December, 1–9, (2003)
30. B. Kuizon, T.L. Melocoton, M. Holloway, S. Ingles, E.W. Fonkalsrud, I.B. Salusky, Infectious and catheter-related complications in pediatric patients treated with peritoneal dialysis at a single institution. Pediatr. Nephrol. 9(1), S12–S17 (1995)
31. P.H. Hien, N.H. Tam, N. Van Xuan, The reversible air dryer SRA: one step to increase the mechanization of post-harvest operations. In International Conference on Crop Harvesting and Processing (p. 10). American Society of Agricultural and Biological Engineers, (2003)
32. P. Kuizon, Rice husk furnace and reversible airflow grain dryer, in *ACIAR Proceedings*, 356–359 (1996)
33. P. H. Hien, N. V Xuan, N. H. Tam, L. V Ban, T. Vinh, Grain Dryers in Vietnam, *Ho Chi Minh City, Vietnam Vietnam Agric. Publ.*, (2000)
34. H.T.A.M. Nguyen, N.V. Xuan, P.H. Hien, Results of reversible air drying research. J. NLU Agric. Sci. Technol. (1), 81–90 (2002)
35. P.H. Hien, N.H. Tam, N. Van Xuan, Study on the reversal timing for the sra reversible dryer. City 11, 12 (2003)
36. C.J. Tado, D.P. Ona, J.E.O. Abon, E. Gagelonia, T.N. Nguyen, Q.V. Le, Development and promotion of the reversible airflow flatbed dryer in the Philippines. Ann. Trop. Res. 37(1), 97–109 (2015)
37. Y. Yuwana, B. Sidebang, "Performance testing of the hybrid solar-biomass dryer for fish drying," Int. J. Mod. Eng. Res., 6(1), 63–68, November (2016)

Chapter 5
Sustainable Drying Techniques for Developing Countries

World hunger crisis is exacerbated by excessive annual food waste. Nearly 1.3 billion tons of food or a third of all food produced globally do not reach the stomach of people as they are being wasted in the food supply chain [1]. It is a matter of grave concern since harvesting food requires a substantial amount of resources such as energy, human labor, water, and time. This wastage of food and resources is unacceptable, considering the fact that nearly a billion people are deprived of nutritional food. Moreover, the proper utilization of natural resources is paramount for mediating greenhouse gas pollution. In other words, reducing food waste will not only eradicate world hunger but also ensure global sustainability.

A universal solution for food waste is improbable since the degree of economic inequality, policy shortcomings, and lack of awareness vary from country to country. In developing countries, a major portion of food is being wasted in the postharvest stage (i.e., on the way to market or at the farm) [2]. On the other hand, more than 40% of wasted food is discarded by consumers. According to the World Health Organization (WHO), inadequate postharvest food processing facilities of the developing countries primarily contribute to the exorbitant amount of food waste. Hence, infrastructures equipped with proper food processing technology can potentially decrease the food waste generation rate [3]. The researchers have been working tirelessly to design and develop effective and efficient ways of drying food in order to increase its shelf life. In Chap. 4, various aspects (i.e., efficiency, effectiveness, and quality of dried food) of existing drying technologies in developing countries has been extensively studied. Most of those technologies use mediocre efficiency with inconclusive effectiveness. The current generation of researchers has developed numerous improved drying techniques through a rigorous scientific study on material science, renewable energy integration, heat, and mass transfer phenomena. Despite various improved drying techniques have their respective merits and features, they are not always suitable for developing countries. In this chapter, the authors will propose and evaluate the performance of improved drying techniques suitable for developing countries based on their energy efficiency and

M. Hasan Masud et al., *Sustainable Food Drying Techniques in Developing Countries: Prospects and Challenges*, https://doi.org/10.1007/978-3-030-42476-3_5

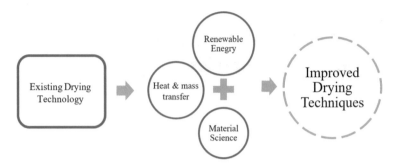

Fig. 5.1 Improvement of drying techniques

economy. Therefore, energy is the primary performance parameter for benchmarking and recommending improved drying technologies for developing countries. Similarly, the quality of the dried products including color, texture, and smell, also needs to be considered. It is established that the quality attributes are highly dependent on process parameters. Rahman et al. investigated the microstructures of different fruits during drying and found that the evolution of microstructure is also dependent on the type of drying and drying parameters [4–6]. Porosity is another physical characteristic that significantly impacts product quality [7, 8].

The purpose of this chapter is to provide extensive discussion on different improved drying technologies and methods in order to recommend them for developing countries. Some dryers are designed for dehydrating only a specific type of food (e.g. grain, powder, fish, etc.). While illustrating the schematic diagrams of various dryers, the authors of this chapter have used an 'apple' to depict most type the food materials. Description of each dryer provides detailed information on the compatible food type that can be dried. A concept map for improving the existing drying technique is depicted in Fig. 5.1.

5.1   Classification of Drying Technique

Drying techniques can be classified based on the various criterion and thermal properties, which were discussed extensively by Mujumdar et al. [9]. In order to evaluate and compare the features of proposed dryers, it is essential to know where they fit in the classification of all available industrial dryers. A general classification of the drying system is presented in Fig. 5.2 [10]. It is important to note that each of these classes can be further divided based on additional properties.

Selection of appropriate drying techniques for a specific application is a challenging task that requires careful attention. Based on their specific characteristics, one dryer can be suitable or unsuitable for a particular operation. Some drying techniques are costly (e.g., freeze dryers) to operate while some others are fundamentally more efficient (e.g., conductive dryers). As such, it is crucial to know the

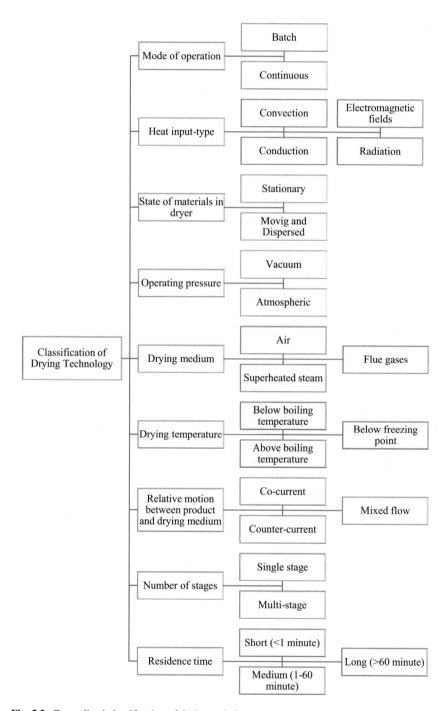

Fig. 5.2 Generalized classification of drying techniques

advantages and disadvantages of the available dryers before recommending one for the developing countries. Typically, dryers with high efficiency have high install-ment and operating costs such as vacuum drying, freeze-drying, dielectric drying, Wyssmont turbo-drying, and Hazemag rapid drying. The authors have collectively decided to exclude these advanced drying techniques from the aforementioned clas-sification as these drying techniques are challenging and costly to manufacture, maintain, and operate. Therefore, these drying techniques may not be suitable for a developing country. Instead, the authors have focused on discussing renewable energy-based improved drying systems with low operating costs. Drying technolo-gies that require low and, in some cases, zero fossil-based energy for its operation are most appropriate for developing countries due to its long-term sustainability. Therefore, the authors believe the following proposed renewable energy-based low-cost drying technologies can be adopted by developing countries to minimize their food waste.

5.2 Improved Solar Drying

Based on the method of harnessing solar energy and storing them, the solar dryer can be classified into different categories. The design of these dryers, their way of heat, and mass transfer phenomena vary broadly between each type of dryer. Solar dryers are inherently economical since it utilizes the sun's abundant solar energy and requires less maintenance as it has fewer moving parts. Over many years of research and development, scientists have developed various models of solar dryers, which are represented in Fig. 5.3. The solar dryer is widely used for its simple design, low initial investment, and moderate efficiency, one major disadvantage of the solar dryer is its ineffectiveness during the night or bad weather. Hence, improved

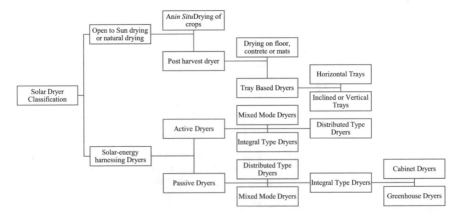

Fig. 5.3 Classification of solar dryers

solar drying model is proposed with the addition of phase change material, which can enhance the efficiency and running time of the existing solar dryer.

Analyzing the thermal condition of the solar dryer with phase change material is a complicated problem, which has been rigorously studied by many researchers, and several mathematical models have been developed over the years [11–15]. The solar collector is the main component of any solar dryer as its performance significantly influences the economy of solar dryers [16, 17]. The most common models consider the temperature of the collector plates and the fluid to be in thermal equilibrium, and they are widely used for thermodynamic optimization. For higher temperature applications, concentrated solar collectors can also be used [18–20]. Solar collector models with phase change materials are developed based on energy flux balance over each component of solar collector. The challenge is to estimate the heat transfer coefficients and numerically solve for the temperature of each component. In order to make the computation simpler, isothermal models are considered. The advantage of these models is that the plates are modelled to be at a mean temperature avoiding the spatial variation. The models evaluate the parameters of the solar dryer with phase change material based on the heat removal factor and the collector flow factor, which combines the solar radiation, air mass flow, and overall loss coefficient. Isothermal assumptions produce valid results as long as the collectors have the appropriate dimension suitable for the respective model. The properties of phase change material can be assessed using the model proposed by Jokisalo et al. [21].

Non-isothermal models are used to solve for the temperature distribution and airflow temperature for the flat plate solar collector independent of the dimension. Numerical models are used to solve these models since it is complicated to obtain an analytical solution. Ong et al. and Forson et al. have developed various types of single-pass flat plate solar collectors with and without glass cover [22, 23].

5.2.1 Phase Change Materials for Thermal Energy Storage in Improved Solar Drying

Phase change materials use their chemical bonds to store and successively release heat. This phenomenon can be explained based on the principle of "Latent" heat storage. When the phase change material (PCM) changes its physical state from a solid to a liquid, or from the liquid state to solid-state, heat transfer occurs within the material as chemical bonds break apart. Initially, the deployed phase change material absorbs heat as their temperature increases. Afterward, when the phase change material reaches its melting point, it consumes a large quantity of heat without increasing its temperature [24]. The PCM maintains a constant temperature until all its mass has melted. The amount of heat stored in the melting process is defined as latent heat. PCM has high thermal storage density and seamless temperature variation. When the surrounding temperature drops below the melting point of the

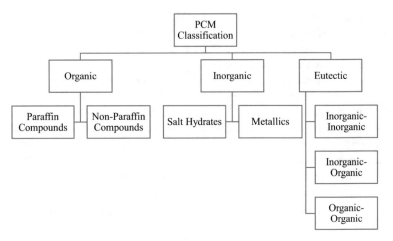

Fig. 5.4 Classification of phase change materials

PCM, the molten PCM starts to solidify and releases the latent heat, which is stored during the melting phase change process. The released heat can be used for drying purposes. As such, PCM integrated solar dyer functions as follows- (i) during the day time in the presence of sun's thermal energy PCM changes its phase from solid to liquid and stores sun's thermal energy as latent heat (ii) when the night falls, and the sun can no longer supply heat, the surrounding temperature around PCM starts to deplete; as a result, the PCM begins to solidify, which releases the stored latent heat to its surrounding (iii) the released heat can be used for drying long after the sun's thermal energy becomes unavailable. Various types of PCM materials are available in the market; a general classification of PCM materials is provided in Fig. 5.4 [25].

Among the PCM materials available in the market, Paraffin is the most commonly used type of PCM. Paraffin wax is structurally a simple chain of n-alkanes CH_3-(CH_2)-CH_3. A large amount of heat is released during the crystallization of the (CH_3) - chain. Paraffins have a characteristic of higher melting point and latent heat of fusion as their chain length increases. Paraffins are reliable, non-toxic, non-corrosive, inexpensive, and can be considered as the heat of fusion material for their ability to operate in a wide range of temperatures. Moreover, paraffin is stable and chemically inert under 500 °C temperature. In Table 5.1, melting point and latent heat of fusion of technical grade, group I and group II paraffin are shown [25].

The performance and operation time of a solar drier can be significantly improved by the integration of a thermal storage medium such as PCM and pebble bed [26–28]. Properties of PCM from the above table can be used to select suitable PCM material for commercial solar dryer applications. Researchers are also investigating the use of nanofluids for thermal storage applications [29]. Various improved solar dryers with thermal storage systems are discussed later in this chapter.

Table 5.1 Thermal properties of commonly used paraffins

Classification	Paraffins/Number of carbon atoms	Melting points (°C)	Latent heat of fusion (kJ/kg)
Technical grade Paraffins	6106	42–44	189
	P166	45–48	210
	5838	48–50	189
	6035	58–60	189
	6403	62–64	189
	6499	66–68	189
Group I	14	5.5	228
	16	16.7	237.1
	18	28	244
	20	36.7	246
Group II	15	10	205
	17	21.7	213
	19	32	222
	21	40.2	200
	22	44	249
	23	47.5	232
	24	50.6	255
	25	49.4	238
	26	56.3	256
	27	58.8	236
	28	61.6	253
	29	63.4	240
	30	65.4	251
	31	68	242
	32	69.5	170
	33	73.9	268
	34	75.9	269

5.2.2   Overview of Different Types of Proposed Improved Solar Dryers

Researchers have employed various solutions to overcome the limitation of the unavailability of solar energy during night and cloudy days. Among those techniques, the integration of a thermal storage system in order to extend their operation time beyond the off-sunshine hours is most promising. An overview of the improved solar drying techniques for developing countries is presented here.

5.2.2.1 V-Corrugated Plate Solar Collector With PCM Dryer

For many years, flat plate solar collectors have been used for solar water heating and solar air conditioning by various industries and domestic consumers. Researchers have developed numerous mathematical models for effective solar flat plate thermal collector design. Subsequently, numerical models have been simulated to optimize the system. In its simplest form, a flat plate solar collector consists of a pump that increases the kinetic energy of air and a thermal collection chamber through which the air flows. When the air flows through the flat plate thermal collector, it takes heat from the hot collector plates. Drying can be achieved via passing the heated air through a food chamber, eventually turning the setup into a flat plate solar dryer. Designing the thermal solar collector is critical for improving the overall efficiency of the system.

The performance of flat plate solar collectors has been improved via the addition of fins, which enhances heat transfer area and the turbulence of air through the duct [30–32]. Additionally, different designs of corrugated surfaces (i.e., V-corrugated) can be added to increase the surface area for efficient heat transfer [17]. Incorporation of thermal storage medium (i.e., phase change materials) with improved flat plate solar collectors can significantly extend the functioning time of a solar drying long after the nightfall. This not only extends the application time of flat plate solar dryer but also complements its efficiency, as observed by Kabeel et al. [33]. Through their experiments, they found that the outlet temperature of the flat plate solar dryer with phase change material was 7.2 °C higher (from inlet temperature) after the 3.5 hours of sunset. Moreover, improved V-corrugated solar dryer equipped with phase change material can potentially have 15–21.3% higher efficiency than conventional flat plate solar collectors without a thermal storage system while operating at a mass flow rate of 0.062 kg/s. A schematic diagram of an improved V-corrugated solar dryer with PCM as the thermal storage medium is shown in Fig. 5.5.

Butler and Troeger have assessed the capacity of a sun-oriented dryer with rock bed storage for drying peanut, an indigenous fruit to many developing countries [34]. The drying experiment was conducted using 4.9 m^3/s air flow rate and continued around 25 hours, which depleted the moisture content of the peanut from 20% (initial moisture content) to storage safe moisture level. From the experiments of Garg et al. [35] and Kabeel et al. [33], performance of low-cost solar thermal storage systems with thermal storage (e.g., solar flat plate air with PCM and augmented integral rock system solar dryer) can be explored. The significance of using PCM on solar air heating and drying is described in some literature. Studies done by Enibe [36] shows the use of PCM for drying medicinal plant. The reports by Mohammed et al. [37], Tyagi et al. [38], and Farid et al. [37] reported the effects of PCM on the performance on solar air heater. Properties of PCMs are classified and reported by Sharma et al. [39].

Fath has investigated the potential of paraffin as PCM using copper cylinders [40]. His findings reveal that when the outlet temperature is 5^0 C higher than the ambient temperature, and the airflow rate is 0.02 kg/s, the daily efficiency can be as high as 63.35%. Mettawee and Assassa have conducted experiments on paraffin, placing the PCM below the absorber plate [41]. The results have shown that with the increase of the PCM layer thickness, the convective heat transfer coefficient can be

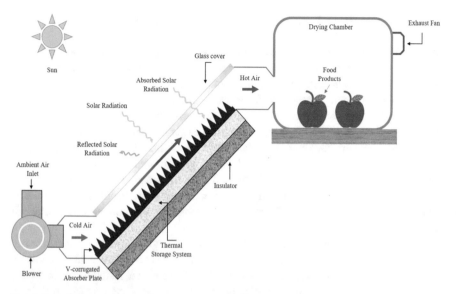

Fig. 5.5 Schematic diagram of improved V-corrugated solar air heater with PCM as a thermal storage system

improved significantly. The thermal performance of a solar collector using hytherm oil and paraffin wax was studied by Tyagi et al. [42]. Paraffin wax is more effective than hytherm oil since it can deliver a higher outlet temperature under the same condition. They have also concluded that the efficiency of a solar collector with thermal energy storage (TES) is 20–53% greater than a setup without TES. Alkilani et al. have investigated PCM freezing time, which has an inversely proportional effect with the mass flow rate of air [43]. The experiment was conducted in an indoor solar air collector equipped with PCM. Operating at a mass flow rate of 0.05 kg/s the PCM was able to reach an extended time of 8 hours. The performance of finned plate and double pass V-corrugated solar collectors were studied by El-Sebaii et al. [44, 45]. They reported that in their experiment when the mass flow rate was set to 0.02 kg/s, conventional flat plate solar air collectors were 9.3–14% less efficient than finned V-corrugated solar collectors [44, 45]. Furthermore, experiments were carried out within the mass flow rate range of 0.009–0.062 kg/s to assess the efficiency of V-corrugated and flat solar collectors [40, 44, 45]. The overall performance of an improved solar dryer with PCM thermal storage can be measured using the following attributes: the regular average efficiency, the rapid thermal efficiency, air temperature difference, thickness of PCM layers, free time of PCM layers, and convective heat transfer coefficient.

Comprehensive studies were done on different designs of solar collectors, including flat plate, V-corrugated, finned, without finned, integrated with and without thermal storage system to assess their efficiencies and effectiveness [23, 33]. It is evident from the literature that the V-corrugated PCM integrated solar thermal collector is the most efficient and has the longest operation time. Therefore, this design can be recommended for the preparation of an improved solar drying system.

5.2.2.2   Solar Dryer With Integrated Rock Storage System

Similar to the principle of a PCM V-corrugated solar dryer, an integrated rock storage solar dryer drives ambient air over an absorber plate equipped with rock thermal storage. The efficiency of the system is improved with the addition of a chimney, which creates upwards natural draft for the incoming air and a plenum chamber for better heating and ventilation. The natural draft created by the chimney eliminates the requirement of a blower, which reduces the operational cost of the system. For a specific rock bed thickness and mass flow rate, where the dryer temperature is significantly higher than the surroundings, it is recommended to use a single or double glass cover in order to limit thermal loss. Irrespective of the mass flowrate, an integrated rock storage solar dryer is effective up to 3:30 pm. The system displays high overall thermal efficiency in contrast to the conventional solar air heater [25].

The schematic diagram of a solar dryer with an integrated rock storage system is shown in Fig. 5.6. This method of solar drying requires no mechanical component, which considerably increases its service life. Ayensu and Asiedu-Bondzie have investigated the drying characteristics of tropical products such as cassava chips, cassava leaves, fish, and pepper using an integrated rock storage solar dryer [46]. They have tested the mass transfer rate, convective heat transfer coefficient, and energy-absorbing capacity of the rock storage system [46]. The results show that it is possible to transfer 118 W/m² power to the drying air when the system is operating at 32 °C. Additionally, the drying time of the above system is nearly half in contrast to the time required for an open-air sun dryer. Throughout the process, drying is accomplished in two stages. In the first stage, 'free' water is removed from the food material, which is followed by the second stage, 'bound' water removal. The efficiency of this improved solar dryer is 22%, where the rock thermal storage system enhances the drying capacity by 1.1 kWh^{-1}.

Tiwari et al. have experimented with wheat crop grain drying using the sensible rock-bed heat storage system. Rocks with physio-chemical specifications of 5–8 cm diameter, specific heat of 0.81 $kJ/kg\ K$, and density of 1750 kg/ml were used in the experiment [47]. The wheat crop drying time with a specific moisture content has been evaluated using the experimental data. It was evident from the data that the temperature fluctuates significantly due to the storage effect. Researchers have drawn the following conclusion from their experimental results:

- During the drying process of the wheat crop, the steady-state condition is achieved after 2 hours for using thermal storage capacity
- For a given temperature, moisture content of the food sample decreases as the operation time progresses
- Rock beds with high thermal capacity will require a longer time to reach steady-state condition, and
- The maximum drying temperature of the food sample can be reduced, utilizing a thermal storage system, which substantially improves the agricultural process of production.

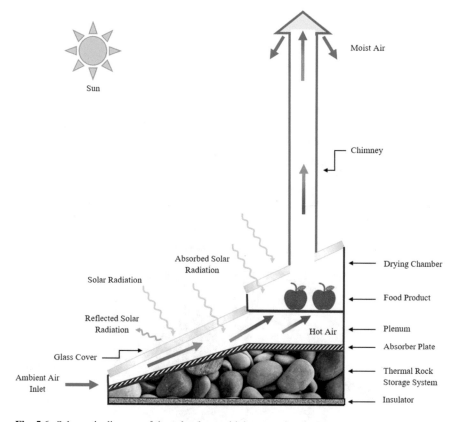

Fig. 5.6 Schematic diagram of the solar dryer with integrated rock storage system

5.2.2.3 Deep Bed Dryer With Drying Bin, Air-Heater and Rock-Bed Storage

Deep bed dryer integrated with bin, air-heater, and rock-bed storage is essentially the combination of a flat plate solar collector and an integrated rock bed storage system with chimney and plenum. This hybrid dryer is created fusing the novel features from each of those two types of dryers. A deep bed dryer has two intricate systems; the first one is an independent solar air heater, and the second one is a rock-bed thermal storage system. In order to accomplish drying, hot air is generated during the sunshine hours using the solar air heating section. In addition to supplying hot air to the drying chamber, a separate solar air heating section is used to supply thermal energy to the rock bed storage system. During the off-sunshine hours when solar air heater is ineffective, the rock bed storage system provides adequate thermal energy to the air to continue the drying process. Furthermore, a chimney is installed to create a natural upwards draft for the incoming hot air, and the plenum is used for efficient heating and ventilation. A schematic diagram of the deep bed dryer integrated with a bin-air-heater-rock-bed storage system is shown in Fig. 5.7.

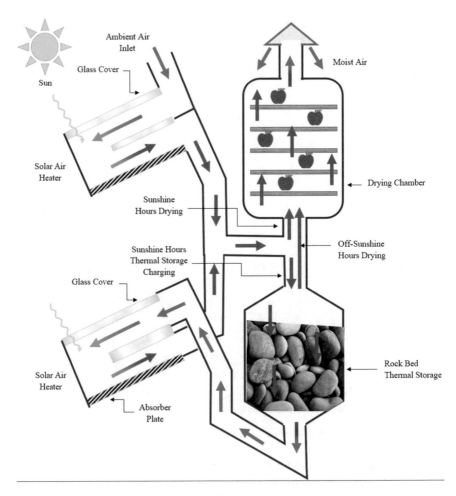

Fig. 5.7 Schematic view of a deep bed dryer with drying bin-cum-air-heater-cum-rock-bed storage

In a study, Chauhan et al. have dried coriander using a deep bed dryer integrated with bin-air-heater-rock-bed with a capacity of 0.5 tons/batch [48]. During the drying process, coriander grains were stacked on top of each other in several thin layers. Energy and mass balance equations have been developed for each component of the system using a finite difference approach to assess its theoretical potential. The results of the simulation show that standalone solar air heater requires almost 27 cumulative hours of sunshine (3 days of sunshine) in order to reduce 16.8% of moisture content from its initial 28.2% (*db*) to 11.4% (*db*).

On the other hand, in order to remove the same amount of moisture in similar conditions, deep bed dryer integrated with bin-air-heater-rock-bed storage system requires 18 hours of sunshine and 13 hours of off-sunshine. In other words, a deep bed dryer integrated with bin-air-heater- rock-bed storage system can accomplish the same result with 31 cumulative hours or two days and two nights. During the simulation, the air mass velocity was taken to be 250 kg/hm^2. It was found that during sunshine hours, air mass velocity has no discernible effect on the reduction rate of moisture content. However, the bed depth of the coriander grain affected the performance of the dryer significantly. Moreover, the drying time of off-sunshine can be reduced to 12 hours if the mass flow velocity is increased to 300 kg/hm^2 from 250 kg/hm^2. Conversely, the off-sunshine time will increase by 1 hour if the mass flow rate is reduced to 200 kg/hm^2 from 250 kg/hm^2. Hence, this principle will allow consumers to utilize the heat stored in rock beds effectively and according to their need for drying agricultural products.

5.2.2.4   Multi-Solar Roof Desiccant Dryer With Desiccant-Arrangement

A multi-solar roof dryer with desiccant arrangement capitalizes on the working principle of solar air heating with an added benefit of humidity control using desiccant material. In its simplest form, the system consists of a multi-storeyed solar air heating chamber that increases the thermal energy of the ambient air and flows it down through the desiccant chamber, where its moisture content is absorbed via desiccant material. Ducts, vents, and necessary pumps are used to maintain the required airflow rate throughout the system. Once the hot air filters through the desiccant chamber, it is mixed with the ambient air in order to produce the desired drying environment. The mixed air is then supplied to the drying chambers containing a bulk volume of agricultural products. A schematic diagram of a multi-solar roof dryer with the desiccant arrangement is shown in Fig. 5.8.

Improved solar dryer with this configuration has been studied by Ziegler et al. [49]. In their experiment, a layer of wheat was used as desiccant material, which

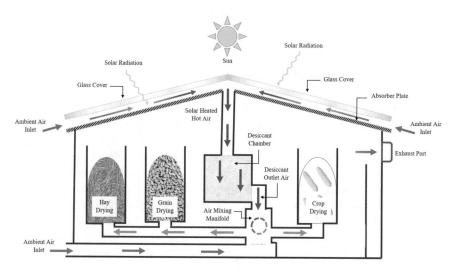

Fig. 5.8 Schematic diagram of a multi-solar roof dryer with the desiccant arrangement

Table 5.2 Properties of various industrial desiccant materials

Desiccant materials	Regeneration temperature (0C)	Volume flow rate of regeneration air (m³/h)	Process air humidity ratio (g/kg)	COP
Silica gel	70–160	600–5400	7–24	0.3–2.7
LiCl	60–120	300	14–19	3–4.8
Natural zeolite	60.8	400	9.5	0.34
Metal silicate	-	250–750	21	1.0–2.5
Alumina charcoal	54.7–68.3	80–120	19.7–24.9	–

was integrated with a deep bed solar air dryer. This type of dryer is highly recommended for drying materials in bulk, such as wood chips, hay, and grains. Since this dryer offers continuous storage drying, with no requirement for fossil fuel combustion, it is incredibly safe and reliable. Lower-income developing countries may use grain as desiccant material; since it is significantly cheaper and immune dust damage. The strategy for maintaining a proper drying environment is to control the mixing process of solar-heated air with ambient air flowing through the desiccant system. Throughout the system's day and night operation, an ideal 65% relative humidity of the drying medium can be ensured. This is critical for developing countries situated in a high humidity climate. In order to prevent the growth of mold, maintaining a low level of humidity inside the drying chamber is imperative.

Furthermore, the desiccant material can be regenerated by using the solar heated air in a low-flow mode without any additional cost or energy. The performance of this improved drying can be upgraded using industrial desiccant material, albeit it will increase the initial investment cost. Properties of different industrial desiccant materials can be seen in Table 5.2 [50].

5.2.2.5   Reversed Solar Absorber With a Thermal Storage System

Reversed solar absorber is an ingenious design that increases the effective heat transfer surface area without enlarging the system dimensions. This is accomplished by a concentrated solar collector dish that focuses the solar rays on the bottom absorber plate of the drying chamber. Therefore, a reversed solar absorber dryer is equipped with two absorber plates. One in the top, which absorbs the direct solar radiation and the other, is on the bottom, which utilizes the reflected solar radiation from the parabolic concentrated reflectors. Since this installment can effectively supply heat from two directions, it has the potential to reduce drying time significantly. Two modes of heat transfer also qualify this dryer to be used for drying large quantity of agricultural products. The operational time and efficiency of the system can be improved with the integration of the thermal storage system. The schematic diagram of a reversed solar absorber with thermal storage is shown in Fig. 5.9.

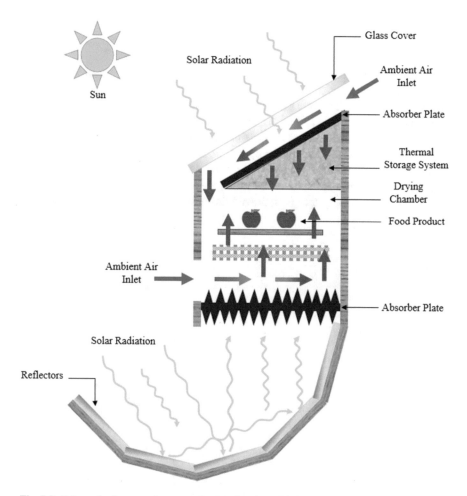

Fig. 5.9 Schematic diagram of a reversed solar absorber with thermal storage

Jain et al. have developed a transient logical model that can be used to evaluate the prospect of this new and improved solar dryer design, equipped with reversed absorber plate and thermal storage [51]. Onions were dried in trays using a packed bed configuration of this setup. The performance of the reversed solar absorber was assessed by analyzing the drying characteristics of onions. The temperature inside the drying chamber is significantly influenced by the packed bed height and air flowing channel width. The fluctuating drying temperature is neutralized by the thermal storage system. Besides, a thermal storage system extends the operational time of a dryer beyond the off-sunshine hours. The mathematical model used in the Jain et al. study can be used for performance evaluation of the reversed absorber type dryer. This type of dryer is recommended for convective solar crop drying with natural airflow.

5.2.2.6 Indirect Type Natural Convection Solar Dryer

Indirect type natural convection solar dryer is a relatively new technology. The system consists of a thermal mass collector-storage system, a convectional solar drying chamber with chimney, and a biomass burner equipped with a flue gas chimney and rectangular plenum. The schematic diagram of an indirect type natural convection solar dryer is shown in Fig. 5.10. This type of dryers function based on the following working principle: (i) food or agricultural products are placed on top of a biomass burner (ii) during the drying operation, moisture from the food material is removed using hot air generated by the biomass and solar air heating (iii) additional heat which dissipates to the surrounding is absorbed the thermal storage system. The efficiency of the system is improved through the utilization of waste heat. In

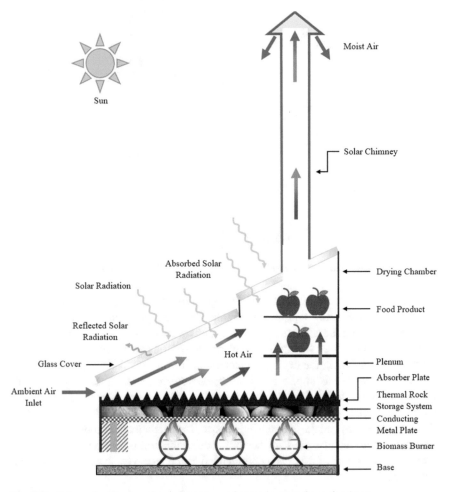

Fig. 5.10 Schematic diagram of an indirect type of natural convection solar dryer

addition to extending the operating time of the dryer during the off-sunshine time, the thermal storage system also minimizes the drying temperature fluctuation.

Indirect natural convection solar dryer has been extensively studied by Madhlopa and Ngwalo [52]. They have designed, prototyped, and evaluated its performance with rigorous experimentation. Using the indirect type natural convection solar dryer integrated with thermal storage and biomass-back up the heating system, they have dried fresh pineapple. The conventional solar dryer can operate only on bright sky days. However, using biomass burners as backup, this improved drying system can overcome that limitation. Results from the study conducted by Madhlopa et al. show that after drying, the moisture content of pineapple was reduced from 66.9% (*db*) to 11% (*db*). The solar-biomass heat transfer model has significantly low drying time due to its capacity for generating high thermal flux.

5.2.2.7   Multi-Tray Solar Crop Dyer

Multi-tray solar crop dyer is a natural convective dryer with improved drying capacity. Using a number of stacked trays, a bulk amount of food material can be dried in one single batch. This technique not only saves time but also reduces the cost of drying. Capitalizing over the principle of solar collector, the air is heated for the purpose of drying with an absorber plate. One or more glass plates are used to reduce thermal losses. Air is recirculated through a multi-pass glass plate to increase the instantaneous efficiency of the system. A natural upwards draft is created using a chimney on top of the system; this reduces the power required to maintain airflow throughout the system. Additionally, the dryer can be equipped with a thermal storage system in order to increase operational performance and reduce drying temperature fluctuation. The schematic diagram of a multi-tray solar crop dryer is shown in Fig. 5.11.

The performance of a multi-tray solar crop dryer has been assessed by Jain et al. through rigorous experimentation and periodical analysis [51]. The experiments were conducted on a multi-tray solar crop dryer equipped with a thermal storage system. Hourly moisture content reduction and drying rate in various trays can be measured using thin-layer drying equations developed by Jain et al. [51]. The results of their experiment show that as drying time progresses, the moisture content of crop decreases. Due to the difference in temperature in separate trays, the drying rate varies from tray to tray.

5.2.3   Comparison Among the Different Improved Solar Dryers

All of the improved solar drying techniques have their unique advantages and specific strength over others. Some dryers can function longer, whereas some dryer process bulk quantity of agricultural products. In order to choose an appropriate drying technique, it is imperative to delineate the effectiveness of each dryer. Based

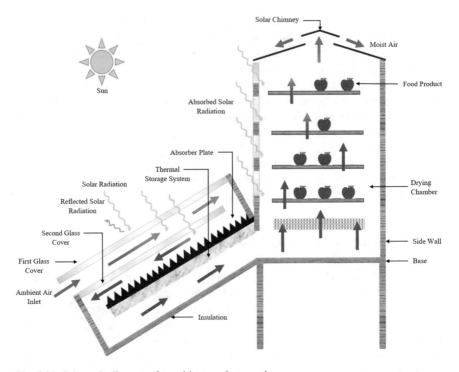

Fig. 5.11 Schematic diagram of a multi-tray solar crop dryer

on the performance analysis, an effectiveness pyramid chart has been constructed in Fig. 5.12.

From Fig. 5.12, it is apparent that the deep bed dryer with drying bin-cum-air-heater-cum-rock-bed storage system is the most effective due to is low drying time and long operational time. This technology is inexpensive and can process/store a large quantity of food material. Its moisture removal rate is also significantly high. An improved design that combines this technology with revered solar absorber can improve the efficiency of the system considerably. Similarly, it is possible to integrate two or more improved solar drying methods with superior performance.

5.3   Geothermal Drying

In order to ensure the availability of nutritious food all year round, especially during droughts season, food products can be processed and preserved via drying. Food drying decelerates the process of wastage and increases their shelf-life. According to literature, geothermal resources with low to medium enthalpy and earth's undisturbed temperature (EUT) below 150 °C are adequate for agricultural drying [53, 54]. In the case of geothermal drying, the heat is obtained from the geothermal well

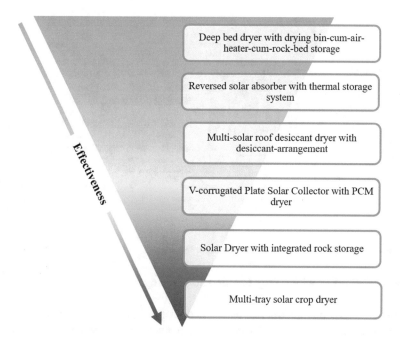

Fig. 5.12 A comparison among the various improved solar drying techniques

streams, hot underground water, and recovered waste heat from a geothermal plant [55]. Most prominent advantages of using geothermal energy for food drying include the low operational cost and abundance of stream and hot water sources [56].

Moreover, this technique is considerably eco-friendlier than fossil fuel and electricity-based drying. An enormous amount of energy is required for food drying, which can be harnessed from geothermal resources. For example, it was found that the former Yugoslav Republic of Macedonia required 136 *kWh/tons* of thermal energy for drying rice [57]. Similarly, Greece required 1450 *kWh/tons* for drying tomato [58]. Low to medium enthalpy geothermal assets with temperatures under 150 $^\circ C$ are utilized on the grounds that have the most noteworthy potential for rural drying applications [53, 54]. Different types of agricultural products, including but not limited to rice, tomatoes, cotton, garlic, chilies, wheat, and onions can be dried using geothermal energy.

5.3.1 Recommended Geothermal Dryer for Food Products

Low-cost sustainable drying technologies can be adopted by tapping into the abundant geothermal energy sources available to certain developing countries. Geothermal resources supply direct thermal energy, which makes geothermal dryers easy to install. A snapshot of the feasible geothermal drying technologies for developing countries is presented here.

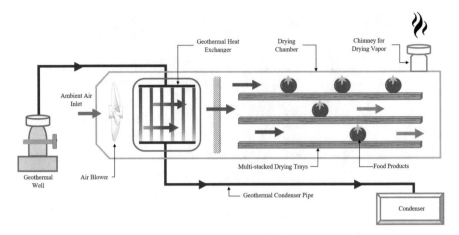

Fig. 5.13 Schematic diagram of a Geothermal dryer

5.3.1.1 Multi-Stacked-Tray Geothermal Drying

A multi-stacked-tray geothermal drying is one of the most conventional geothermal drying mechanism. In this setup, geothermal hot water is pumped from underground and passed through a cross-flow heat exchanger. Ambient air is heated using the thermal energy of hot water. Multi-stacked-tray drying is also installed with a geothermal power plant so that it can use the residual waste heat for air heating. The heated air is then passed through the drying chamber, designed with multi-stacked-trays. Food materials are placed on various trays.

Geothermal drying using a multi-stacked-tray mechanism can be found in Nea Kessani, Xanthi, Greece, which started its operation in 2001 [59]. This drying plant has 14 m rectangular dryers with a dimension of 2 m heights and 1 m width tunnel. They dried tomato slices using 59 °C hot water retrieved from the geothermal resource. Around 7 kg of raw tomatoes can be dried in a single batch of 25 trays, which takes around 45 minutes to accomplish. The setup used in Greece for tomato drying is shown in Fig. 5.13. The tomatoes are then further processed by immersing them in olive oil. This plant has successfully produced and transported 4 tons of high-quality dried tomato during its first year of operation.

Additionally, in 1992, a pilot project for cotton pre-drying was also established in Nea Kessani, Xanthi, Greece [60]. The cotton drying plant was also designed based on the principle of multi-stacked-tray geothermal drying. Results from the pilot project show that a significant amount of cotton can be dried using an improved geothermal water-based tower drier. A schematic diagram of the geothermal cotton dryer used in Greece is shown in Fig. 5.13.

5.3.1.2 Geothermal Cabinet Dryer

Geothermal cabinet dryer utilizes recovered waste heat from a geothermal power plant as its primary source of thermal energy. Typically, waste heat is recuperated as hot water, which is then passed through a counter flow heat exchanger in order to heat the air. The heated air is subsequently passed through a closed cabinet where the food products are placed. The cabinet system is beneficial since it enables the user to control the temperature and humidity inside the drying chamber. Appropriate drying conditions for each specific type of agricultural product ensures that better-quality dried foodstuff can be achieved.

Additionally, the geothermal cabinet dryer system requires low drying time. It is also more energy-efficient than multi-stacked-tray geothermal drying. This type of dryer can be used to dry garlic and chilies, which are essential to the agricultural economy of South East Asian countries (i.e., Thailand, Bangladesh, and India). In Thailand, geothermal cabinet dryers have a dimension of 2.1 × 2.4 × 2.1 m (width × length × height). Each dryer can produce 220 kg of dried garlic and 450 kg of dried chilies. Garlic and chilies ware washed and placed on the 36 plates inside a two-compartment configuration. Recovered waste heat from a nearby geothermal power plant is used to supply thermal energy to the dryer. Boiling water with a temperature of 80 °C is passed through a counter-flow heat exchanger, which has a dimension of 100 × 500 × 300 mm (width × length × height). A steady airflow of 1 kg/second blows through the heat exchanger and enters the 10.5 m³ drying load. The required temperatures for drying chilies and garlic are 70 °C and 50 °C, respectively. Geothermal cabinet dryers of Thailand require around 46 hours for drying chilies at 1 kg/second mass flow rate and 94 hours for drying garlic at 0.04 kg/second mass flow rate. The energy requirement for evaporating 1 kg of water from chilies and garlic are 13.3 MJ/kg H_2O and 1.5 MJ/kg H_2O, respectively. Two significant advantages of this category of the dryer are that it has low operating cost and are functional in all type of weather conditions. The schematic diagram of a geothermal cabinet dryer is shown in Fig. 5.14.

5.3.1.3 Geothermal Convective Dryer for the Rice Drying

A geothermal convective dryer is built in Kotchanya, Macedonia, for rice drying [59]. Hot water from a nearby geothermal well is utilized for direct heating. Ambient air with a relative humidity of 60% and a temperature of 15 °C is circulated through a water-air heat exchanger (WAHX). The WAHX increases the temperature of ambient air up to 35 °C. Hot water is used as heat exchanger fluid for the WAHX setup. The inlet temperature of water in the WAHX is 75 °C and the outlet temperature is 50 °C. When the air is heated to the desired temperature, it is flown to a drying chamber containing rice samples. The air moves downwards through the drier at a constant velocity. The drying capacity of this dryer is 10 tons/hour equipped with a 1360 kW heating capacity. In order to prevent the rice from cracking, the temperature inside the dryer is kept below 40 °C. Using this dryer, the moisture content of

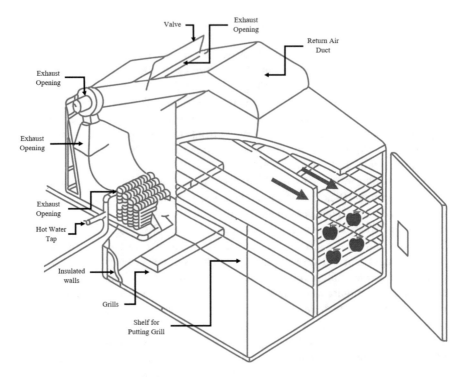

Fig. 5.14 Geothermal cabinet dryer

rice can be reduced from 20 (%wb) to 14 (%wb). The rice is air-cooled after the completion of the drying process. The efficiency of the convective dryer can be improved by integrating this system with other heating technologies such as industrial dryers and greenhouse agriculture farms. The resultant dryer has the potential to compete with the performance of a liquid fuel industrial dryer [57]. Schematic diagram of a convective geothermal rice drying setup is shown in Fig. 5.15.

5.3.1.4 Geothermal Fruit Drying

Lund and Rangel designed and fabricated a setup for drying fruits using geothermal energy in 1995 [61]. This setup was installed and tested in the geothermal field of Los Azufres, Mexico. The system was developed using the following dimensions 1.35 m width, 4 m length, and 2.3 m height. The structure was made of timber ceiling concrete walls and reinforced concrete floors. This dryer had a capacity to dry 1 ton of fruits per cycle, which can be placed inside the dryer using two containers, each with 30 trays. When the geothermal hot water flow system runs at 0.03 kg/ seconds, it consumes energy at a rate of 10 kJ/second. The optimum drying temperature inside the dryer was 60 °C. Under these ideal conditions, the dryer was able to reduce the moisture content of fruit samples from 80 (%wb) to 20 (%wb) within

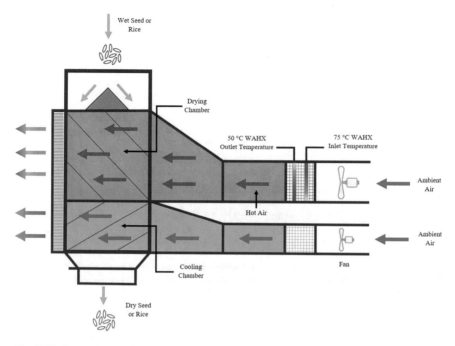

Fig. 5.15 Convective geothermal rice drying

24 hours of drying time. The schematic diagram of a geothermal fruit drying system is revealed in Fig. 5.16.

The efficiency of the system can be improved using cascading, which will also reduce the operational and maintenance cost of the geothermal resource. A local community of Eburru, Kenya, dries their agricultural products like tobacco, maize, and pyrethrum using recovered heat from a geothermal stream [62].

5.3.1.5 Geothermal Tube-Bank Heat Exchanger Dryer

Indonesia, with its enormous potential of geothermal resources, has developed the geothermal tube-bank heat exchanger drying system. This technology has enabled the people of Indonesia to dry various types of food products including, tea, beans, coffee berries, fishes, and rough rice [63]. In the West Java Kamojang geothermal field, a geothermal dryer with this special configuration has been installed for grain and beans drying. Steams generated by the geothermal well has a temperature of 160 °C, which is used for heating the ambient air. The heated air is then supplied to the dryer. The ingenuity of this system is that air is flown through a geothermal tube-bank heat exchanger where it is heated to a desirable temperature. The intricate design of the tube-bank heat exchanger improves system efficiency. Once the heated air reaches the appropriate temperature, it is blown into the drying chamber, which consists of four trays. The geothermal tube-bank heat exchanger of this system has

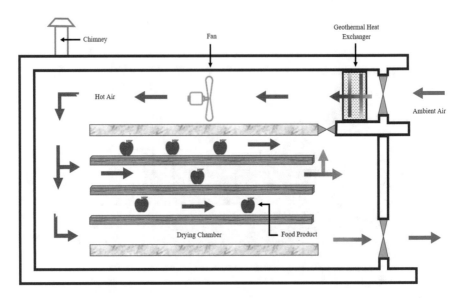

Fig. 5.16 Fruit drying using geothermal energy Mexico

a heat transfer rate of 1000 W. The ideal air velocity range of this dryer is 4–9 m/s, and the drying temperature range is 45–60 °C. The system's drying time is significantly affected by the moisture present in the raw material. The schematic diagram of a geothermal tube-bank heat exchanger dryer is shown in Fig. 5.17.

5.3.1.6 Deep-Bed and Conveyor Type Geothermal Dryer

Deep-bed (batch dryer) and conveyor type geothermal dryer are two independent drying system which can be categorized as an advanced geothermal dryer. Albeit these two systems are expensive, they have considerably high efficiency and capacity. An industrial-scale conveyor geothermal dryer has been installed in western Nevada, the USA, for onion and garlic drying. A continuous conveyor that is used in that facility has a dimension of 60 m length and 2.8 m width. The system has a staggering capacity of drying 3000–4300 kg of wet onions per hour. This is an output rate unmatched by any other type of dryer. Moreover, the moisture removal rate is also significantly high for this type of dryer, which is evident by its ability to produce 700 kg/hour dried onions by depleting its moisture content from 85 (%wb) to only 4 (%wb) within 24 hours [64]. Conveyor type geothermal dryer for fish drying is shown in Fig. 5.18.

Every year grain drying consumes an overwhelming amount of fossil-based energy, which can be reduced through the integration of renewable energy in the drying industries. Deep-bed drying is one such technology that utilizes geothermal energy for efficient grain drying. This system consists of a fan or a blower, which provides the much-needed kinetic energy to the air so that it can flow through the

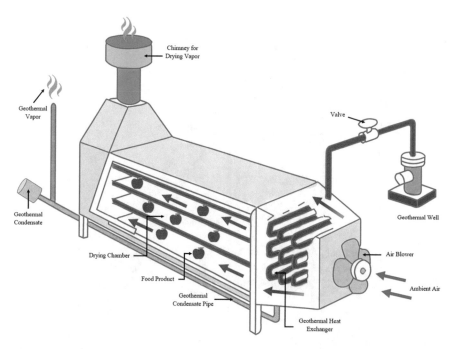

Fig. 5.17 Geothermal tube-bank heat exchanger dryer for beans

geothermal heat exchanger. In this process, the air is heated to an adequate drying temperature, which is then uniformly distributed through the perforated floor of the drying chamber. The hot air temperature of this system can be adjusted by regulating the hot water flow through the geothermal heat exchanger. While the drying temperature for some grains can be as high as 90 °C, for adequate drying, the system operates in a moderate temperature range of 50–60 °C with a relative humidity of 40% [63]. For instance, the required drying temperature for coffee berries is 50–60 °C. However, as mentioned previously, for drying rice, the temperature is kept below 40 °C in order to prevent cracking. After a successful drying cycle, the moisture content is lowered to 12–13 (%wb), which prevents spoilage and mold growth. The schematic diagram of a deep-bed geothermal grain dryer is shown in Fig. 5.19.

5.3.1.7   Rack Tunnel Geothermal Dryer

In Iceland, an industrial scale rack tunnel geothermal dryer is operational for drying fish and seaweed. For over 35 years, Iceland has been using its geothermal resources for indoor drying. Among the dried products, small fish, cod backbones, stockfish, and salted fish are most common. Rack tunnel geothermal dryer is used by nearly 20 companies all across Iceland for drying fish products [59]. In its simplest form, a rack tunnel geothermal dryer is similar to a multi-stacked tray geothermal dryer,

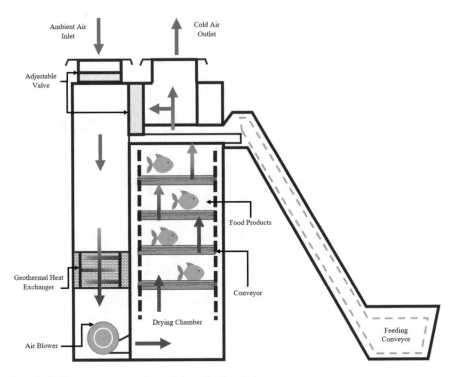

Fig. 5.18 Conveyor type geothermal dryer for fish drying

except it has a hot air recirculation system, as represented in Fig. 5.20. Since rack
tunnel geothermal dryers are used for drying bulk quantity, hot air is recirculated in
order to save energy and properly recuperate the thermal energy inside the hot air.
Chimneys are installed inside the dryer to create a natural draft, which ensures continuous air circulation throughout the dryer. Geothermal hot water and steam are the
primary source of heat, which are used to increase the temperature of ambient air.

A typical rack tunnel geothermal dryer has a capacity ranging from 2000–4000
tons per batch, which is considerably high compared to other geothermal dryers.
Drying of fish is completed in two steps; in the first step, moisture is reduced from
80 to 55 (%wb) by drying the fish products for 24–40 hours in a 20–26 °C temperature range [56]. In the second stage, the products are dried continuously for three
days using 20–26 °C temperature air, which lowers the moisture content down to
15%. The emerging dried pet food industry with an annual global demand of 500
tones can also use this method of drying [56, 65].

5.3.2 Comparison Among Different Geothermal Food Dryers

According to the report of Lund, Freeston, and Boyd (2010), geothermal resources
are used for heating and air conditioning purposes in over 34 countries including,
the Russian Federation, Italy, China, Hungry, and China [66]. Many of these

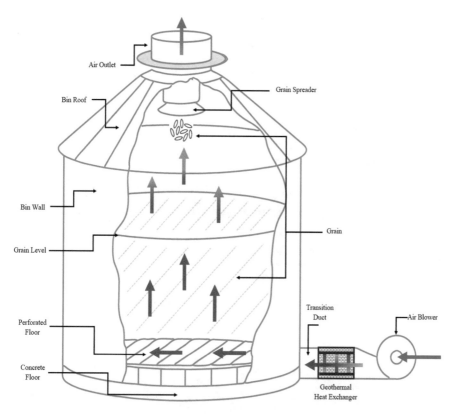

Fig. 5.19 Deep-bed geothermal grain dryer

countries are classified as developing countries; hence, it is highly recommended that these countries invest in geothermal energy-based drying technologies. Fruits and vegetables grown in these regions are appropriate for processing in geothermal dryers. Duffield and Sass (2003) have found that the fuel cost for drying can be reduced by 80% and operating cost by 8% if the industries use geothermal energy instead of fossil fuel [59]. Not to mention, this will have a positive impact on the environment [67]. The transition towards renewable energy will ensure cleaner water and pollution-free air for the next generation. Figure 5.21 delineates the performance of various types of geothermal dryers based on initial investment cost and their respective effectiveness.

5.4 Hybrid Geothermal-Solar Drying

The hybrid geothermal PCM flat plate solar collector is a combination of two independent systems that work in synch with each other [205]. Initially, the ambient air is being filtered in order to prevent bacterial growth and particulate material accumulation inside the geothermal earth to air heat exchanger. The filtered ambient air

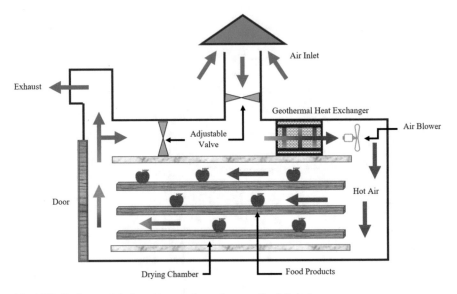

Fig. 5.20 Rack tunnel drying using geothermal energy for fish drying

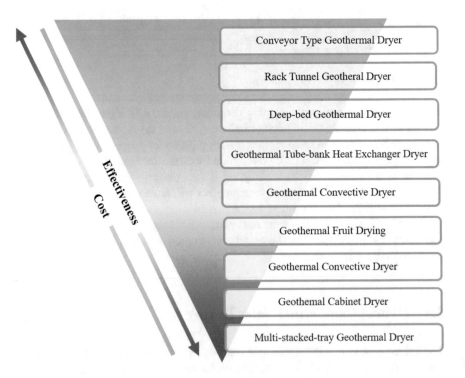

Fig. 5.21 A comparison among the various improved geothermal drying techniques

is pushed down to the insulated flow pipe deep inside the earth at a depth of required temperature gradient based on the geological location using a high lift closed impeller multi-stage radial flow centrifugal pump.

The air flows through the geothermal earth to air heat exchanger with low discharge. During its transmission through the heat exchanger pipe network, the air is heated to a specific temperature above the ambient air temperature, which is then circulated inside the PCM solar flat plate solar collector. The preheated air is subsequently heated to the required drying temperature and pushed into the drying chamber using a blower to maintain the necessary mass flow rate. A model of the hybrid geothermal PCM Flat Plate Solar Collector is shown in Fig. 5.22, where the geothermal heat pump is adapted using the earth to air heat exchanger (EAHX) [205]. Since the temperature is higher than the ambient temperature of the output air of the EAHX, it would significantly increase the efficiency of the system in comparison with the traditional solar dryer. The PCM Flat Plate Solar Collector (PA-FPSC) consists of three layers of the heat transfer medium. Firstly, the layer between the glass and collector plate, where the glass layer functions as a greenhouse and helps to increase the temperature of the compartment between the glass layer and solar heat absorber. Then comes the solar heat absorber (SHA); this plate will absorb the solar and glass plate radiation, which will increase its surface temperature. Beneath the SHA plate, there will be a compartment where PCM will be kept with a thickness of 20 cm. The PCM will store energy in the form of latent heat of melting that is obtained from the solar radiation through the solar heat absorber plate. There will also be fins in the lower portion of the SHA plates to increase the heat transfer between the PCM and SHA plate. There will be an insulation layer in the lower part

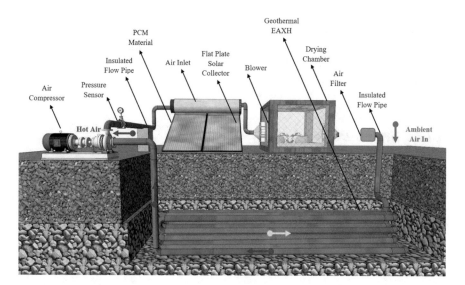

Fig. 5.22 Concept design of hybrid geothermal PCM flat plate solar collector

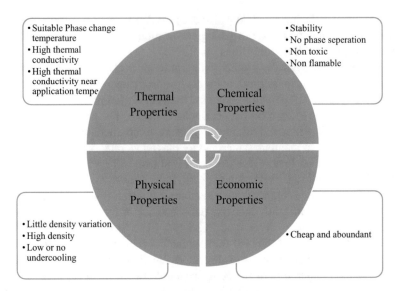

Fig. 5.23 Some essential characteristics of PCM

of the compartment so that the heat doesn't radiate or conduct on any other direction accept the SHA plate.

The selection of PCM is vital to the functionality of the system. Energy storage materials should have some distinctive properties, which are shown in Fig. 5.23 [68].

Paraffin C_{33} can be used for the following reasons, commercial paraffin waxes are cheap, and they have moderate thermal storage densities. Additionally, paraffin waxes have a wide range of melting temperatures, and they are chemically inert and stable with no phase segregation [37].

However, it has a low thermal conductivity, which limits its application. But this limitation can be minimized by the use of metallic fillers, metal matrix structures, finned tubes, and aluminum shavings [69]. The properties of paraffin wax C_{33} is shown in Fig. 5.24 [70]:

5.4.1 System Description

A heat and mass transfer model of PCM flat plate solar collector is shown in Fig. 5.25, using a rectangular cross-section [205]. On the top layer, solar radiation is partially absorbed by the glass plate, which raises its temperature. Heat is then transferred by convection from glass plate to air coming from the geothermal EAHX that is forced to flow inside the channel. A portion of the solar heat flux is reflected from the top surface of the glass plate via radiation. Part of the heat absorbed by the

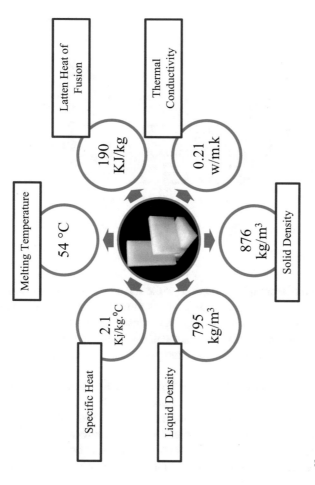

Fig. 5.24 Properties of paraffin wax

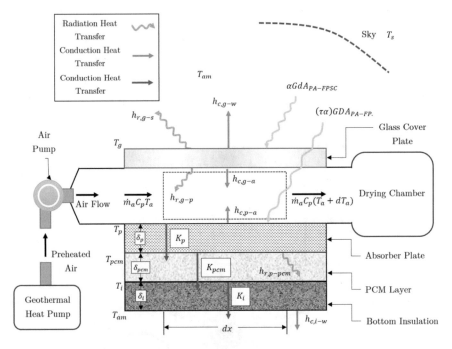

Fig. 5.25 Schematic diagram of hybrid PCM flat plate solar collector

glass is transferred (through convection) to the surrounding ambient air passing over the glass plate. A portion of the heat is radiated from the glass plate to the copper absorber plate below, which also receives transmitted solar radiation. Since the absorber plate is painted with a mat black color, it will absorb maximum solar radiation. Subsequently, it will release the energy as heat in the flowing air and the PCM layer below the absorber plate.

The layer of paraffin wax PCM receives heat energy from the absorber plate in the form of conduction and radiation. Since the study deals with solid PCM, the convection within the layer of paraffin wax can be ignored. The insulated bottom layer is considered adiabatic. Solar radiation exerts a uniform heat flux on the glass plate and absorber plate. Since there exists convection between the plates and airflow, the temperature of components, the fluid, the glass cover plate, and the absorber plate, must vary along with the direction of flow. Hence, the absorber plate and the glass plate are considered non-isothermal. To compute the temperature distribution in each layer of the PA-FPSC, energy flux balance equations were used parallel to the flow direction. The EAHX is made of copper pipe, which is of uniform cross-section with very small thickness, i.e., the thermal resistance is negligible. The schematic diagram of the EAHX heat transfer phenomena is represented in Fig. 5.26.

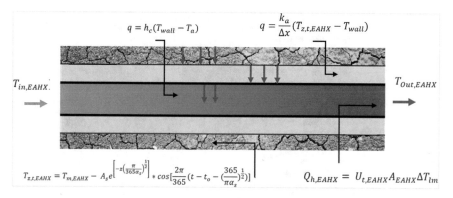

Fig. 5.26 Schematic diagram of earth to air heat exchanger

5.4.2 Feasibility Analysis

5.4.2.1 Performance Evaluation

The mathematical models for the geothermal heat pump simulated by earth to the air heat exchanger and the PCM flat plate solar collector were solved separately to determine the effect of different parameters on the output of the system. The output temperature of both sub-systems was considered to be the primary output parameter since the drying condition specifically depends on the output air temperature. The objective was to determine the efficiency at which the hybrid system can maintain the required drying temperature.

It can also be seen that because of the presence of phase change material, the hybrid system can maintain the optimum drying temperature longer than any other conventional flat plate solar collector. This is a clear indication that the hybrid system is more efficient and produces more useful thermal energy than the traditional standalone flat plate solar collectors (FPSC).

In addition, from numerical evaluation, it is evident that the hybrid geothermal PA-FPSC can sustain an average of 19.16 °C higher than ambient air outlet temperature over 24 hours a day. In comparison, 13.34 °C higher than the average outlet temperature of conventional FPSC over 24 hours a day [205].

Figure 5.27 describes the instantaneous efficiencies of various FPSC drying systems [205]. Since the geothermal system is highly efficient, it is seen that the instantaneous efficiency of hybrid geothermal PA-FPSC is often higher than any other system. However, when the PCM material completely uses all its latent heat and turns into solid, the efficiency of the hybrid system drops significantly. In the morning, when the PCM material starts to heat up, the efficiency of the system begins to decrease, provided that heat is being stored inside the system. After the activation of the solar energy during noon, the efficiency again rises, and at dawn, with the activation of PCM material, the efficiency drastically increases. Later, when the PCM material starts to freeze, the efficiency of the system drops gradually only to rise again when the PCM has achieved a stable temperature.

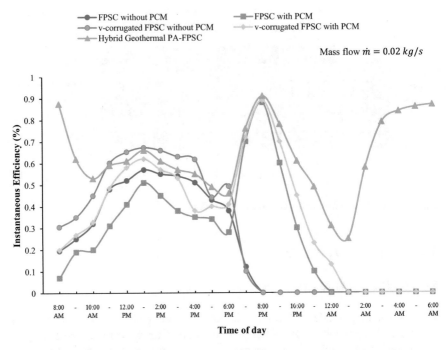

Fig. 5.27 Comparison of the instantaneous efficiencies among various types of FPSC with hybrid geothermal PA-FPSC

The result of the numerical solution supports the hypothesis made by the authors and conclusively provides that the hybrid geothermal PA-FPSC has higher efficiency than the conventional solar collector.

5.4.2.2 Industrial Aspects in Developing Countries

The increasing energy demand in this rapidly industrialized world would not meet up if the transformation from dependence on fossil fuels to the renewable energy source is done quickly. Geothermal energy is viable, environment friendly, and has more potential to serve as an unlimited energy source. Studies on the application of heat energy for cloth drying [71], textile drying [72], biomass drying [73] have been conducted. Several studies on the food drying process have also been performed. The results of the proposed hybrid dryer suggest that the existing food drying industry can be improved by using the Hybrid Geothermal PCM Flat Plate Solar Collector. The heat energy used for conventional drying of food originates from fossil fuel, which is harmful to the environment and increases the carbon footprint of post-harvest food production [74]. The relation between the conventional energy-burning and CO_2 emission is proven beyond doubt. Therefore, it is necessary to be conservative about using fossil fuel food drying as it would emit more CO_2 in the

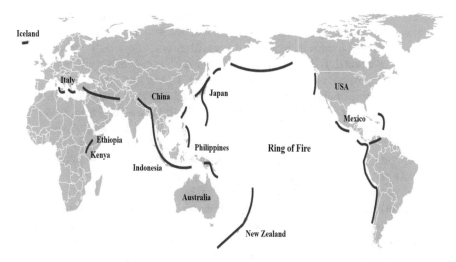

Fig. 5.28   Major location of geothermal potential around the world

environment [75]. However, the proposed food drying process has almost no negative impact on the environment. Thus, this scheme does not emit additional CO_2 in the environment, but itself is a clean alternative to the food drying system using fossil fuels.

From Fig. 5.28, it is clear that numerous developing countries such as Bangladesh, India, Kenya, Ethiopia, Chile, Paraguay, Peru have significant geothermal potential, which is essential for the adaption of the proposed hybrid geothermal PCM flat plate solar collector. This system can operate nearly 20 hours based on renewable energy, and as a result, this has long term economic viability. Moreover, the reduction of food waste and the protection of the environment would give this system a decisive advantage over the traditional drying techniques practiced in developing countries.

5.5   Heat Pump Drying

Heat pump drying technologies have proven to be considerably energy efficient when they are combined with conventional drying processes. Heat pump dryer has the ability to recuperate wasted thermal energy from exhaust gases. Additionally, heat pump dryers have superior control over the humidity and temperature of the drying medium. Researchers, through rigorous experimentation, have identified the optimal drying conditions for various agricultural food products and fisheries. When the food samples are dried under ideal condition, it produces better-quality dried products. MacArthur et al. have emphasized the necessity of optimizing the system parameters of a heat pump drying system to make it more energy-efficient [76, 77]. Universally, any drying technology that usages convection as its primary mode of heat and mass transfer (with or without a secondary mode of heat transfer) can be integrated with a heat pump design. Dryers that are most commonly used in conjunction with heat pumps include tray dryers, batch shelf fluid beds, rotary dryers, and kilns for wood drying [78]. Heat pump operation is not compatible with dryers that require enormous amounts of drying air such as, spray or flash dryers.

For developing countries with high humidity ambient air, a heat pump dryer can be very effective, since it can adequately control the humidity during drying. In developing countries, the food spoilage rate increases during the rainy season due to highly moist air; therefore, heat pump dryer can effectively remedy the crisis. Through proper humidity control, a heat pump dryer can prevent agricultural products from spoilage.

Heat pump drying not only yields improved quality dried food but also reduces the amount of operating energy in contrast to a conventional drying system. This has been observed by Rossi et al. through their attempts to dry onions using heat pump dryers [79]. The efficiency of the heat pump drier is increased while drying foods with high water content. This is possible because high latent heat energy is transferred to the evaporator when the drying air absorbs more moisture content. As a result, higher heat recovery is possible, and lower energy is required to operate the compressor for sensible air heating.

In accordance with the trend of improving product quality and efficiency, many researchers have studied various features of heat pumps. Results from their experiment have ensured the steady growth of heat-pump drying technologies. Major researches on the performance of heat pump agricultural drying are summarized in Table 5.3.

Industries have been using a heat pump as an efficient energy recovery system for many years. When the air stream passes through a condenser, heat pumps can transform air's latent heat of vapor condensation into sensible heat. This feature, along with the fact that it can operate in conjunction with existing drying systems, make heat pumps a highly sought-after technology. Because of these advantages, heat pump dryers have been used for decades in wood kilns, dehumidifying air, and producing quality lumbers [91]. The generalized classification of the heat pump dryer is provided in Table 5.4. Most of these heat pump dryers have been invented over the last two decades, and a few of them are still in the early development phases.

5.5.1 Working Principle

A heat pump (HP) dryer basically operates on the principle of the vapor compression refrigeration cycle. The primary components of a heat pump dryer include a condenser, an evaporator, a compressor, and an expansion valve. The condenser and evaporator function as heat exchangers. The schematic diagram of an air to air heat pump cycle is shown in Fig. 5.29. The schematic diagrams of pressure-enthalpy and temperature-enthalpy of the HP system are shown in Figs. 5.30 and 5.31, respectively. The working principle of a heat pump cycle is described below.

Table 5.3 Research on heat pump drying

Source	Year	Location	Application	References
Nassikas et al.	1992	Greece	Paper	[80]
Rossi et al	1992	Brazil	Vegetable	[79]
Meyer and Greyvenstein	1992	South Africa	Grains	[81]
Strommen and karmmer	1994	Norway	Marine products, i.e., fish	[82]
Mason and Balrcom	1993	Australia	Macadamia nuts	[83]
Carrington et al. and Sun et al.	1996	New Zealand	Timber and wood drying	[84]
Prasertsan et al.	1997	Thailand	Banana drying	[85]
Chou et al. and Chua et al.	1998	Singapore	Agricultural and marine products	[86]
Pal et al.	2008	India	Green sweet Pepper	[87]
Aktas et al.	2009	Turkey	Apple	[88]
Hii et al.	2012	Malaysia	Cocoa beans	[89]
Shi et al.	2013	China	Yacon slices	[90]

Table 5.4 Classification of heat pump dryer

Criterion	Types
Processing mode of the dryer	1. Batch dryer
	2. Continuous dryer
Number of drying stages	1. Single drying stage
	2. Multiple drying stage
Product temperature	1. Above freezing point
	2. Below freezing point
Number of stages of heat pump	1. Single-stage heat pump dryer
	2. Multiple stage heat pump dryer
Auxiliary heat input	1. Convection
	2. Conduction
	3. Radio-frequency
	4. Microwave
	5. Infrared
Heat pump dryer operation	1. Intermittent operation
	2. Cyclic operation
	3. Continuous operation

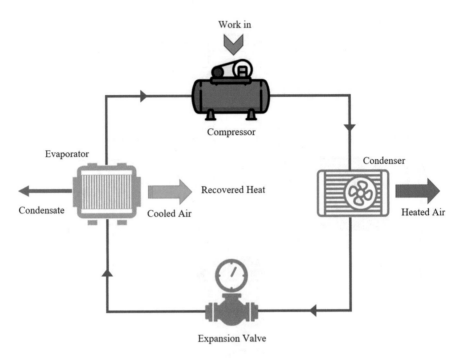

Fig. 5.29 Schematic diagram of an air-to-air heat pump cycle

Fig. 5.30 Pressure–
enthalpy diagram of the
heat pump cycle

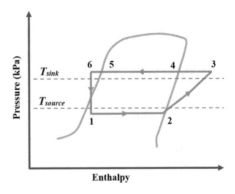

Fig. 5.31 Temperature–
entropy diagram of the
heat pump cycle

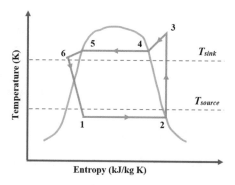

The process can be started from the evaporator, where the air is dehumidified and cooled. Referring to Figs. 5.30 and 5.31, when refrigerant moves from point 1 to 2, heat is absorbed from the air by the refrigerant, which enables its phase change from the vapor-liquid mixture to vapor. During this process, pressure and temperature of refrigerant remain constant. At point 3, the refrigerant vapor enters the compressor through the suction line. At point 4, the compressor increases the pressure of refrigerant vapor to the pressure level present in the condenser. As the vapor pressure rises, the condensing and boiling temperatures of the refrigerant is increased to a level higher than the surrounding temperature. In this stage, the superheated vapor is obtained.

Following the compression stage, the superheated refrigerant vapor is condensed when it passes through a heat exchanger (condenser), releasing heat to the environment. Superheated vapor is condensed in two consecutive stages. In the first stage, de-superheating turns superheated vapor into saturated vapor, and in the subsequent step, the saturated vapor is directed through the condenser. Inside the condenser, the refrigerant is subjected to a two-phase condensation, which transforms it from vapor to liquid phase. During the condensation process, heat is dissipated to the surrounding air.

Vapor refrigerant may undergo additional sub-cooling from point 5 to 6 using another heat exchanger as it exits the condenser. Sub-cooling has two profound advantages. One, sensible air heating can recover additional heat. Two, flashing can be prevented as the refrigerant pressure drops in the throttling device.

Once the condensing process is completed, liquid refrigerant is expanded using a throttling device like a capillary tube, valve, or orifice plate. The primary function of the throttling valve is to reduce the refrigerant liquid pressure and maintain a boiling temperature below the heat source. The entire cycle is repeated once the expansion process is completed, and the refrigerant is redirected to the evaporator.

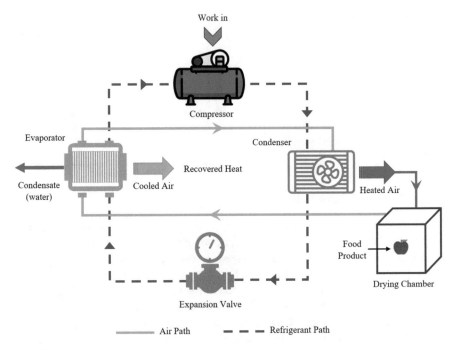

Fig. 5.32 Schematic representation of heat pump drying system

5.5.2 *Mathematical Model*

Mathematical model of a heat pump dryer is developed to assess its performance and efficiency. The schematic diagram of a heat pump dryer equipped with refrigeration components is shown in Fig. 5.32. Hot air enters the drying chamber at point 1 and absorbs moisture from food products. Moist air leaves the dryer at point 2 and passes through the evaporator coil. There are two types of evaporation systems. In the first type, the direct expansion coil method in which refrigerant is subjected to a two-phase change from liquid to vapor to accomplish dehumidification and cooling of air. In the second type, called the chilled water method, controlled flow of chilled water cools and dehumidifies the air. Dehumidification is achieved by first cooling the air sensibly to its dew point when it progresses from point 2 to 3. Afterward, evaporator absorbs the latent heat of vaporization in order to boil the refrigerant. Heat recovered in this process is pumped to the condenser unit. From point 5 to 1, the dehumidified and cooled air absorbs the heat from the condenser. During this step, the air is sensibly heated to the desired temperature.

The coefficient of performance (COP) of a heat pump dryer is given by:

$$COP_{carnot} = \frac{Useful\ heat\ output}{Power\ input} \tag{5.1}$$

Carnot efficiency or the maximum theoretical efficiency of a heat pump dryer can be defined as:

$$COP_{carnot} = \frac{T_{condenser}}{T_{condenser} - T_{evaporator}}$$
(5.2)

COP_{carnot} can be used to assess the refrigeration effects of a heat pump. Generally, a heat pump has an actual efficiency of 40 to 50% of the theoretical Carnot efficiency [92, 93]. Specific moisture extraction rate (SMER) is used to evaluate the performance of a heat pump dryer. Following equation can be used to determine SMER:

$$SMER = \frac{amount\ of\ water\ evaporated\ from\ the\ dryer}{energy\ input}\ kg\,/\,kWh$$
(5.3)

Additionally, specific energy consumption (SEC), which is reciprocal of SMER, can be used to delineate the energy efficiencies of different dryers. The relation between SEC and COP [93] is given by the following equations:

$$SEC = \frac{W}{G_M} = \frac{\dfrac{Q_{ev}}{COP_A - 1}}{M_a(\omega_1 - \omega_2)} = \frac{h_1 - h_2}{(COP_A - 1)(\omega_1 - \omega_2)}$$
(5.4)

5.5.3   Recommended Heat Pump Dryers for Food Products

The high efficiency and low operating cost of a heat pump dryers make it ideal for large scale operation. Consequently, its improved performance and better retention of food quality reduce the payback period by generating substantial profit. An overview of the feasible heat pump drying technologies for developing countries is presented here.

5.5.3.1   Conventional Heat Pump Dryers

Single Stage Perpendicular and Parallel Flow Heat Pump

Heat pump dryers can be designed following two different modes of operation, one is parallel flow, and the other is perpendicular flow. A parallel flow batch heat pump dryer is displayed in Fig. 5.33. Food products are placed inside the dryers by organizing them in an array of trays. Once the food products are dried to a desirable quality, they are removed from the dryer. The hot drying air flows parallel to the

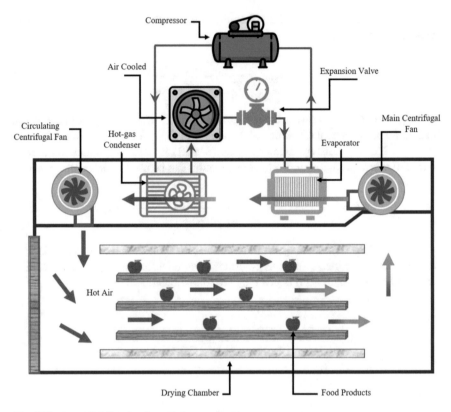

Fig. 5.33 A parallel flow batch mode heat pump dryer

product surface. Although batch dryers require high labor costs, they are suitable for small scale food processing operations.

Figure 5.34 shows the heat pump dryer with perpendicular flow configuration. Similar to the parallel flow heat pump dryer, food samples are positioned on an organized array of trays. Hot air is flown perpendicularly over the trays as hot air enters the drying chamber from top and leaves from the bottom of the dryer.

Another common type of batch is called continuous mode, in which food products are placed on a moving conveyor belt. Using a gear controlling mechanism, conveyor belt speed can be adjusted. Advantages of the continuous systems include the quick loading and unloading of the agricultural products and the requirement of low manpower. Based on the required loading capacity and drying characteristics, the suitable drying method should be selected.

Multistage Heat Pump

In order to improve dryer capacity and reduce drying time, a multistage heat pump dryer is designed. This eliminates the limitations faced by the single-stage vapor compression cycle, where a single evaporator is utilized for dehumidifying and

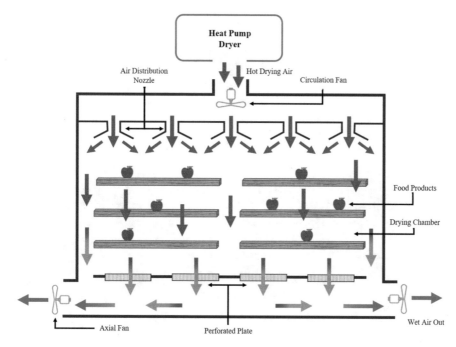

Fig. 5.34 A perpendicular flow batch mode HP dryer

cooling. This imposes a mechanical constraint on the amount of latent heat of vaporization that can be recovered for a specific heat transfer area. Besides, single-stage heat pump dryers cannot produce variations in temperature and humidity for independent drying chambers containing different food materials. However, these issues can be solved using a multistage heat pump. Schematic diagram of a two-stage heat pump dryer is shown in Fig. 5.35.

A two-stage heat pump dryer not only has superior energy efficiency but also has the facility to generate hot air at a different range of temperature and humidity. This type of dryer is recommended for bulk agricultural product drying.

5.5.3.2   Advanced Heat Pump Dryers

Owing to the growing demand for well-controlled and low-temperature drying conditions, heat pump dryers are receiving considerable attention. Recent years have seen a proliferation of new heat pump drying technologies, which are capable of providing high-quality dried food. Freeze-drying techniques are used to process high-value products with extreme heat sensitivity. However, as freeze-drying is overwhelmingly expensive and challenging to operate [94], many industries are taking an interest in heat pump drying as an alternative for freeze-drying.

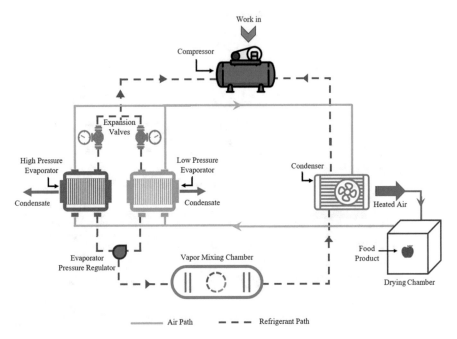

Fig. 5.35 Two-stage heat pump system coupled with a drying chamber

Heat pump dryer is critical for the drying process that requires precise control over temperature and humidity. Heat pump dryers are extremely reliable for drying heat-sensitive food products with low drying temperature. The drying temperature of a heat pump dryer can be meticulously adjusted between 20–60 °C. It is possible to create a freeze-drying environment using a modified heat pump dryer with improved controlling accessories [95]. Heat pump drying can provide improved quality of dried food products due to its ability to maintain optimum drying conditions. Chua et al. have found that agricultural dried product quality can be improved in a two-stage heat pump dryer, which can operate in pre-selected cyclic temperature schedules [96]. The color change can be minimized by up to 87% via modulating the temperature-time variation. Similarly, the degradation of food can be reduced by 20%.

Higher heat recovery is possible, and lower energy is required to operate the compressor for sensible air heating. Therefore, it can enhance the performance of heat pump dryers. A snapshot of the advanced heat pump drying technologies is presented in the following subsections.

Fluidized Bed Heat Pump Dryer

In a fluidized bed heat pump dryer, products with high moisture content entered the dryer through the inlet, and finished dried products are discharged through outlet ducts. The drying temperature can be adjusted within the required range by modulating the condenser power. Subsequently, the air humidity can be controlled by

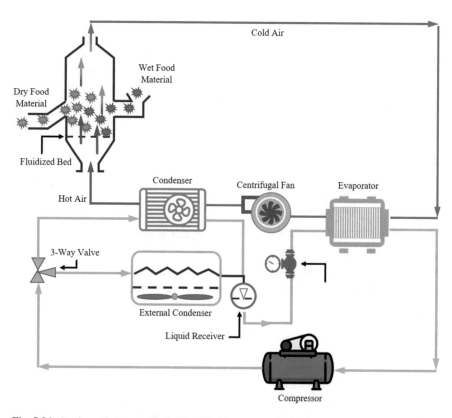

Fig. 5.36 A schematic layout of a fluidized bed heat pump dryer

maintaining compressor capacity using the motor speed frequency control mechanism. Alves-Filho and Strfmmen et al. were able to maintain a drying air temperature range of 20–60 °C and a humidity range of 20–90% [97]. Schematic diagram of a fluidized bed heat pump dryer is shown in Fig. 5.36. Under this controlled drying environment, heat-sensitive foods can be dried effectively. In a fluidized bed heat pump dryer, it is possible to sequence convective and freeze-drying operations. Freeze-drying offers negligible shrinkage of a food sample; however, it has very slow drying rate. Hence, the sequenced application of convective and freeze-drying can improve the drying rate while maintaining a low shrinkage of the food sample. Therefore, advanced fluidized bed heat pump dryer combines both processes where freeze-drying occurs at 58 °C, which is followed by convective drying at 30 °C. Through this mechanism, food qualities such as color, texture, test, porosity, strength, and rehydration rates can be maintained at the desired level [78]. Results of the experiment conducted by Norwegian Institute of Technology (NTNU) confirm that fluidized bed heat pump dryer is extremely effective for drying heat sensitive food materials and pharmaceutical products. Although this method produces premium quality dried products, it is considerably expensive. The incremental drying cost can be offset by the high market price of products.

Fluidized bed heat pump dryers are widely used for drying pharmaceutical products, agricultural crops, and foods with granular solids. Fluidized bed heat pump dryer provides high quality while drying powders in the range of 502000 mm [98]. The efficiency of this dryer and finished product quality it produces is better than the conventional tunnel, conveyor, continuous tray, and rotary type dryers. Fluidized bed heat pump dryer enjoys the following advantages:

1. High heat and mass transfer rates can be observed owing to improved gas-particle contact
2. Flow area is smaller
3. Thermal efficiency is higher
4. Requires low maintenance and capital cost
5. It is easy to operate

However, the fluidized bed heat pump has some limitations like:

1. Consumes high power
2. It is susceptible to attrition, granulation, and agglomeration
3. In case the feed is too wet the system lacks a process of de-fluidization

Solar-Assisted Heat Pump Drying With Energy Storage System

The efficiency of a heat pump dryer can be improved by integrating it with a solar heating system. This is recommended for developing countries with high solar irradiation. This hybrid system is recommended for high-temperature drying. Instead of adopting the conventional mechanism of auxiliary heating using fossil fuel, harnessing the thermal energy of the sun is considerably eco-friendlier.

Additionally, this will significantly reduce the carbon-foot-print of the system. Furthermore, storing solar energy with the aid of phase change material and using it to pre-heat the drying air will lower the operational cost of the heat pump dryer. Subsequently, it will increase the range of drying temperature enabling it for high-temperature drying. Since both solar and heat pump system complements each other, it will make the hybrid system more flexible. Using an integrated thermal storage system, solar energy can be used during off-sunshine hours. The hybrid system, due to its higher efficiency, can reduce the drying time significantly.

A schematic diagram of a solar-assisted heat pump dryer is shown in Fig. 5.37. Main components alongside the standalone heat pump dryer include blower, solar collectors, air-valves, pipes, and thermal storage tanks. The drying parameters can be regulated by modulating the airflow with closed, partially open, or fully opened valves.

A solar-assisted heat pump dryer has the following advantages:

1. This technology is energy efficient due to its ability to operate partially on renewable energy
2. The process is environment-friendly
3. It has a simple control mechanism

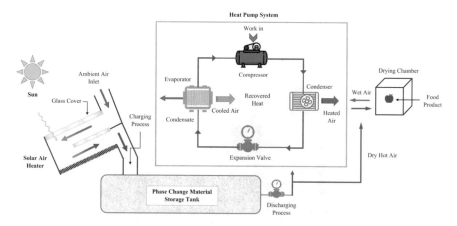

Fig. 5.37 A solar-assisted heat pump system incorporating energy storage system

4. Provides excellent flexibility over-drying parameters
5. Drying can be accomplished at a higher temperature

However, solar-assisted heat pump dryer has the following limitations:

1. High initial capital is required for purchasing the additional solar equipment, including solar panels, storage tanks, valves, and blowers.
2. This system is economically feasible and practical for regions with high annual sunshine time, usually greater than 2600 hours and excellent solar irradiation above 6×10^6 KJ/m^2.
3. The performance and amount of stored solar energy are highly influenced by the weather condition. The solar-assisted section will be useless during the rainy reason.

5.5.4   Performance Analysis

Heat pump drying is accomplished by converting the latent heat of condensation into sensible heat and in the process, increasing the ambient air temperature to the desired level. A higher value of SMER indicates that a dryer has higher efficiency. From the literature, it is evident that drying efficiency and SMER of heat pump dryer is considerably high. Various performance parameters of different dryers are compared with heat pump drying in Fig. 5.38. Additionally, the various cost associated with those dryers is also presented critically.

In summary, a heat pump dryer is an excellent option when product quality is paramount. This process has the capacity for improved drying environment regulation and enhanced efficiency.

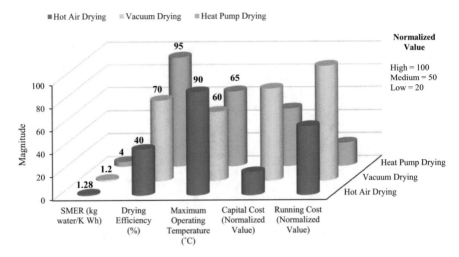

Fig. 5.38 Comparison of heat pump drying performance parameters with other drying systems

5.6 Desiccant Based Drying

The desiccant based dryer is a relatively new low-temperature drying technology with a significant prospect. Equipped with improved air control mechanism, desiccant based dryer can accomplish drying at low-temperature, nearly retaining the original color and texture of food. The current low-temperature drying technologies used in industries are slow and fail to produce uniform drying quality. The improved desiccant-based dryer can be used to solve those issues since it has fast drying time alongside with precise humidity and temperature control mechanism, which causes minimum damage to the food while drying. Researchers have found that vegetables dried using this process significantly retained their vitamin content and fresh color.

5.6.1 Desiccant Drying Principle

A desiccant dryer consists of a dehumidifier equipped with a desiccant wheel. Silica gel is commonly used as a desiccant material [99]. The desiccant wheel has three sections; in the first section, the desiccant material absorbs moisture from the air and supplies the dry air to the other side of the wheel containing the drying chamber. The second section, known as the recovery zone, heats up and blows out the wet air. The third section is called the heat collection zone, which releases heat from desiccant material to the ambient air. As the wheel rotates, the cycle is repeated, constantly supplying hot air to the drying chamber. This process automatically recharges the desiccant wheel [100]. Nagaya et al. have designed and constructed a desiccant dryer with a heater and blower capacity of 1450 W and 50 m³/h of airflow,

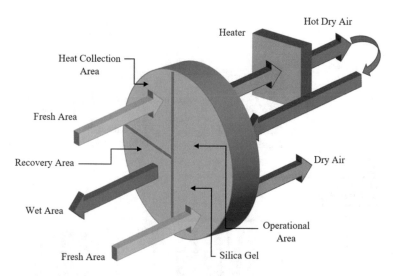

Fig. 5.39 Principle of desiccant dehumidification system

respectively [101]. The dehydrator is rated at 1550 W and 200 V AC supply. Principle of desiccant dehumidification system is shown in Fig. 5.39. The post-process absolute humidity of air is proportional to the absolute suction humidity. The humidity decrease rate of this type of dryer is about 40% [101]. The dehumidification rate considerably high even at low suction humidity and increases with the rise of suction humidity. Research has shown that 800–1100 gh^{-1} dehumidification rate can be achieved using a desiccant dryer with an operation time of 14 hours [101]. This is sufficient dehumidification capacity for a food drying [102].

5.6.2   Desiccant Based Food Drying System

Improved quality of dried food with near original size and color can be processed using a desiccant based dryer. Schematic diagram of a desiccant based dryer is shown in Fig. 5.40. Agricultural products are placed on top of racks. The drying chamber is constructed using clear acrylic plates. When the dryer is in operation, the hot air enters from the bottom of the dryer and blows up through the food and exits the dryer from the top outlet. The temperature inside the drying chamber does not usually exceed 49 °C so that the food texture, color, and vitamin of food can be preserved at its maximum level. Through exhaustive temperature and airflow control, the drying rate is improved, and uniform drying is ensured. Temperature is controlled using fans mounted on the top of the drying chamber. Dry air is heated using an external heater, which is turned on and off using a control mechanism to maintain constant drying conditions inside the drying chamber.

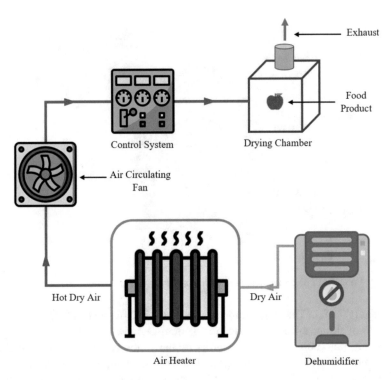

Fig. 5.40 Desiccant based food drying system

5.6.3   Performance Analysis

The change in drying parameters of a desiccant based dryer is determined through rigorous experimentations. Key parameters such as drying time, operating temperature, percentage of humidity reduction, and dried food quality are represented in Fig. 5.41. From the figure, it is evident that although the drying time is high this drying technique produces outstanding quality of dried food. Nagaya et al. have found that when drying is started with a temperature of 28 °C, and after six hours, the dryer temperature rises to 47 °C [101]. Similarly, at the start of drying, relative humidity inside the drying chamber was 50%, after 6 hours, which was reduced to 20% [101]. After 6 hours of drying, food on each tray lost approximately 95 g of water. Additional measurements were made to determine the change in color, texture, and vitamin. The results of that measurement conclude that a significant amount of vitamin of the fresh food product was maintained. Moreover, changes in color and texture were minimum.

This dryer is recommended for drying eggplant, butterbur, carrot, spinach, and cabbage. Although a desiccant based dryer is moderately expensive, it is almost 12 times faster than conventional sun drying.

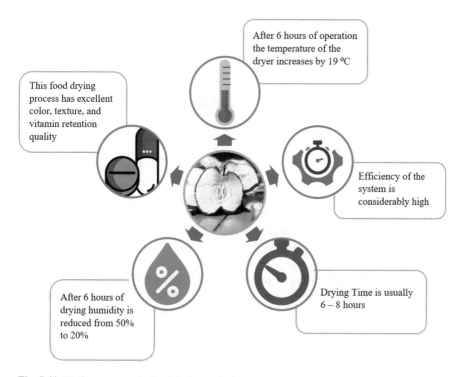

Fig. 5.41 Performance analysis of desiccant drying

5.7 Improved Biomass Drying

An improved biomass dryer can operate on the principle of continuous direct-heat rotary drying, flash drying, or continuous fluidized bed drying. In an improved biomass dryer, the bulk of agricultural food is dried when they come in contact with hot air generated by biomass combustion. The feeds are continuously supplied in rotating cylinders or conveyor belt. Improved biomass drying is a relatively new concept, and further study is needed to assess its performance and feasibility. Generally, organic waste, agricultural residue, animal manure, and sewage sludge are used as biomass feed [103, 104]. The scope of this book allows us to provide introductory knowledge and share the general concepts associated with improved biomass drying.

5.7.1 Recommended Improved Biomass Drying for Food Products

Biomass combustion is a primitive technology for thermal energy production. Albeit it is inexpensive, the efficiency and performance of conventional biomass dryers are matters of concern. The energy efficiency of biomass dryers can be improved via advanced designs and enhanced heat and mass transfer mechanisms. An overview of the improved biomass drying technologies for developing countries is presented here.

5.7.1.1 Continuous Direct-Heat Rotary Biomass Dryers

In this type of setup, a biomass furnace operates in conjunction with a continuous direct-heat rotary dryer. The primary source of thermal energy is produced by the combustion of biomass [105]. Two types of continuous dryers can be selected to design an improved biomass dryer. The first one is an indirect-heat rotary dryer, where heat transfer occurs indirectly. The second one is a direct-indirect rotary dryer, which is essentially a hybrid configuration of direct and indirect heat transfer. Albeit, the design of the direct-indirect rotary dryer is exceptionally complex; it has superior heat loss prevention mechanism as the hot gas does not come in contact with the outer walls [106].

For storage and packaging, dried food products are cooled using air-swept rotary coolers. There may be a dedicated cooling system which can operate concurrently with the drying mechanism. Biomass rotary dryers with a dedicated cooling system can be used for wet sugar processing. The biomass rotary dryer can be further classified into parallel, counter-flow, and cross-flow dryer. As this dryer rotates slowly, it is recommended for drying friable food materials. Schematic diagram of a biomass rotary dryer is shown in Fig. 5.42.

Biomass rotary dryer has the potential to dry any type of food materials except for slurries, solutions, and pastes. The drying efficiency of the system does not depend on the variation of load since this system can generate a substantial amount of thermal energy. This type of dryer is very easy to operate and also highly recommended for industries with minimum supervision. However, biomass rotary dryers can be costly to install, and maintenance costs can be 10% of the investment cost per annum [106].

5.7.1.2 Based on the Principle of Flash Drying

An improved biomass dryer based on the flash drying mechanism should have a biomass combustion chamber, an airflow regulation system, a long duct for high-velocity gas flow, and a feeder, which will disperse and add solid particulates in the gas stream. Additionally, a cyclone collector may be used to recover solids from the gases at the end of drying. Schematic diagram of an improved biomass single-stage flash dryer is shown in Fig. 5.43. In order to improve the efficiency, food particle size should be small, and moisture diffusion should be uniform. Solid feeders such as dispersion mills and venturi sections may be used to supply food particles inside the dryer. For gas-solid separation, cyclones are recommended, which can be installed at a low price. In order to ensure maximum dust recovery, a wet scrubber or bag collector can also be installed with the dryer.

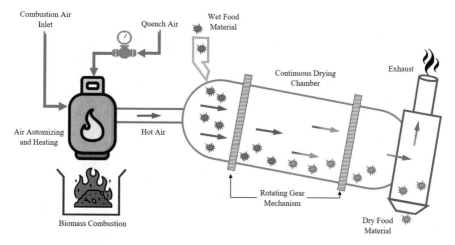

Fig. 5.42 Improved biomass continuous direct-heat rotary drying

Using biomass flash dryer, a maximum of 1–2 mm size particles can be dried [106]. Contrary to the rotary type dryer, the functionality of a flash dryer severely depends on the loading condition. Since overloaded materials cannot be supplied inside the dryer with high feed rates, the flash drying system will fail to operate. Biomass flash drying system would also require more attention than other types of dryers.

5.7.1.3   Continuous Fluid-Bed Biomass Dryer

Biomass continuous fluid-bed dryer will have two main components - a biomass combustion chamber to supply thermal energy to the system and a continuous feeding mechanism that will keep the wet food particle in a fluidized state which will be dried by coming in contact with the hot air. The moisture content of the feed material should be low since the drying efficiency depends on fluidization. Schematic diagram of an improved biomass drying based on the principle of the continuous fluid-bed dryer is shown in Fig. 5.44.

Due to plug-flow characteristics of the continuous fluid-bed dryer, incrustation and improper transportation occur inside the drying chamber. This phenomenon is caused by the small bed height of this type of dryer. In order to solve this issue, the following steps can be implemented (i) additional air should be supplied in the feed section (ii) a transport effect should be created by redirecting blown air and (iii) vibrating the dryer. Vibrating equipment should not reach a temperature of over 300 °C. The air temperature for stationary equipment should be limited to 800 °C [106].

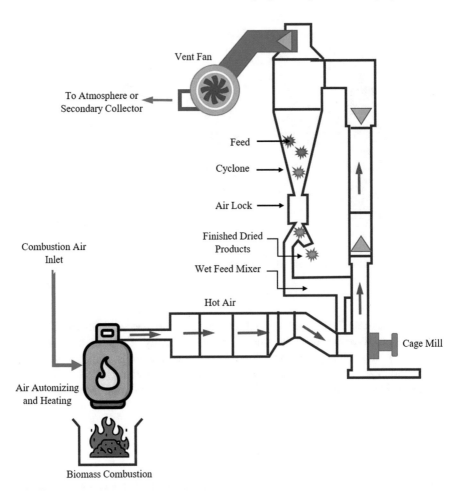

Fig. 5.43 Improved biomass single-stage flash dryer

5.7.2 Comparison Among the Various Methods for Designing Improved Biomass Dryers

In terms of performance, fluid-bed drying falls in between flash drying and direct-heat rotary drying. Direct-heat rotary drying is not influenced by load variation, whereas flash drying is extremely sensitive to loading conditions. Load variation has a moderate effect on fluid-bed drying. Residence time for direct-heat rotary drying, fluid bed drying, and flash drying are 30 min, several minutes, and a few seconds, respectively. Considerable feed disintegration occurs in flash drying, moderate disintegration occurs in fluid-bed drying, and minimum disintegration is observed in rotary drying.

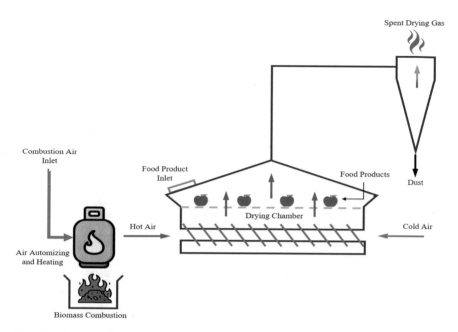

Fig. 5.44 Improved biomass drying based on the principle of continuous fluid-bed dryer

5.8 Drying Using Waste Heat

Despite numerous benefits of drying, it requires a considerable amount of energy during its operation [106]. As a result, the limited fossil-based energy resource has proven inefficient to continuously supply the necessary energy for drying. Conventional drying processes usages biomass, electricity, or fossil fuel in order to generate hot air [107]. However, the high price of non-renewable energy makes it economically unfavorable. In contrast, an astounding amount of heat is wasted from various types of generators, engines, and heating elements typically used in various industries. Researchers, therefore, investigated the potential for developing a drying mechanism based on the waste heat sources [108, 109]. A summary of research works on the utilization and application of waste heat are shown in Table 5.5 [110].

From Table 5.8, it is apparent that the waste heat can be attained from various sources like air-conditioner or refrigerator, IC engine, power plant, cement kiln, brick kiln, and conventional incinerator. These sources dissipate a staggering amount of heat energy to its surroundings [206]. This otherwise wasted heat increases the entropy and caused environmental pollution. A mathematical model is developed by Masud et al. to assess the potential of a waste-heat food drying system [108]. Considering the underlying physics of heat, momentum, and mass transport, the model has been simulated and fabricated for rigorous evaluation. Optimization of the system was initiated to reduce the operating and installment cost. Moreover, the industrial aspect of a waste heat food dryer has been discussed briefly. Utilizing waste heat was a primary source of air heating to accomplish drying. The available temperature of different waste sources is discussed below.

Table 5.5 Different applications of waste heat

Available technology		Outcome	Reference
Direct electrical conversion devices	Thermoelectric generation	Electrical energy	[111]
	Thermionic generation	Produce electricity that can power vehicle auxiliary loads and accessories	[112]
	Exhaust gas heat recovery system (EGHR)	Warm the cabin interior as well as the engine that raises the operating temperature quickly and also boost the fuel economy – Especially in the colder months	[113]
Generating power via mechanical work	Rankine cycle	Produce additional power in the steam generator	[114, 115]
		The coal-to-methanol process with CO_2 capture combined organic Rankine cycle (ORC) for waste heat recovery	
	Stirling cycle	Mechanical propulsion, heating and cooling the electrical generation systems	[116]
	Vapour absorption refrigeration cycle	Heat is provided to a generator which generates the refrigerant vapors that finally facilitates the total refrigeration system.	[117, 118]
	Turbo-compounding	Reduce the brake specific fuel consumption (BSFC) compared to other WHR systems.	[119]
Drying	Biomass drying	Waste heat from a process industry plant (100 MW output) was utilized as the heat source for drying.	[73]
Heating	Preheating	Primary sources of waste heat in industrial facilities include exhaust gases from fossil fuel-fired furnaces, boilers, and process heating equipment may readily be utilized to preheat combustion air, boiler feed water, and process loads.	[120]

5.8.1   Available Sources of Waste Heat

Inefficient tools and thermodynamic restrictions on equipment and methods cause waste heat. For instance, the need for many systems to reject heat as a by-product of their operation is fundamental to the laws of thermodynamics. However, there is a dire lack of information on the source of the enormous heat losses in different sectors and processes and the nature of varying waste heat sources (e.g., the waste heat quality and chemical composition). Knowledge of these factors is critical in determining the feasibility and extent of opportunity for waste heat recovery [110]. Nevertheless, the different sources of waste heat with its exhaust temperature are shown in Table 5.6 [121–123].

Table 5.6 Several sources of waste heat

Sources of waste heat	Exhaust temperature ($^{\circ}C$)
Diesel engine	300–500
Gas turbine	370–540
Ceramic kiln	200–300
Cement kiln	200–350/300–450
Container glass melting	160–200/140–160
Boiler	230
Food industry	164
Conventional incinerator	760
Nuclear power plant	300
Textile industry	60–180
Pharmaceutical industry	330

5.8.2 Proposed Hybrid Drying System

A drying setup based on waste heat and solar PV system was designed. The setup consists of a Photovoltaic cell (provides electricity to run the blower which circulates the air into the heat exchanger system), heat exchanger (works on the principle of the counter-flow heat exchanger), control unit (controls mass flow rate), air circulating unit, battery and a drying chamber (where finally the hot air is circulated), as shown in Fig. 5.45 [206].

The food samples were sliced into appropriate geometric shape and dried using hot air. Using a heat exchanger system, ambient air (low temperature) receives thermal energy from the high-temperature exhaust gas coming out of the waste heat source. When the ambient air passes through the heat exchanger, it absorbs the residual thermal energy of the engine in the form of exhaust flue gas. In this process, the ambient air temperature increases while the exhaust flue gas temperature decreases. The heat exchanger system should be adequately insulated to avoid heat loss. Additionally, the heat exchanger should be airtight so that toxic exhaust gas does not come in contact with the hot drying air.

5.8.3 Mathematical Modelling

The design of the proposed waste heat drying system included a tube-in-tube heat exchanger consist of two tubes- an inner and an outer coiled together, which prevents thermal fatigue, increases efficiency, reduces the overall size, and can work in high temperature, high pressure, and low flow applications. It is primarily used to transfer heat between the liquid and suction lines of the refrigeration plant and has

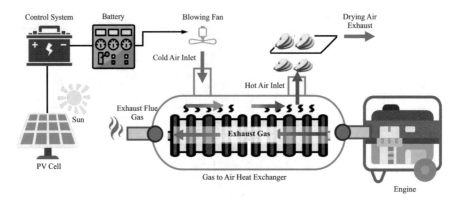

Fig. 5.45 Waste heat convective drying

a high heat transfer coefficient and low thermal resistance in contrast to conventional heat exchangers. Besides, its high thermal performance also allows for compact design, which reduces the operating and fabrication cost. Moreover, its simple design and easy application make it highly recommended for industrial applications, and it can be seamlessly integrated with the waste heat sources.

Although a more effective heat recovery system such as optimized RC network is available, the authors considered the application of heat exchanger so that the proposed system can be incorporated in developing countries (see Fig. 5.46). The proposed design has opted primarily for rural farmers who are not technologically skilled or knowledgeable; hence, a simple design is preferred, albeit having inferior heat recovery potential with high fouling factors and low maintenance [206].

5.8.4 Feasibility Analysis

5.8.4.1 Large-Scale System and Industrial Outcomes

A significant amount of energy required for water migration during drying, which can be provided in the form of heat with maintaining different processing conditions. Under moderate conditions, approximately 12.71 MJ/kg energy is required for a convective dryer to remove water from a specific food [124, 206]. However, with the variation of process conditions and food samples, the energy needed in the drying process varies considerably [125, 126]. The required energy is provided from conventional sources of energies, including electricity, natural gas, and biomass [127, 128].

Dumping of waste heat as exhaust gases from different sources, including IC engine, power plant, Gas turbine, Ceramic kiln, Cement kiln, brick kiln, and conventional incinerator, causes both energy loss and global warming. Therefore,

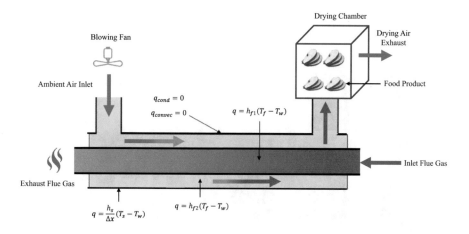

Fig. 5.46 Boundary conditions of the air heating system

utilization of this waste heat on a large scale and industrial level can reduce the dependency on the national grid for energy requirement to a great extent and save the environment from pollution and dry tons of food without taking any conventional energy sources. An example of a Large-scale application of waste heat in the drying of food materials is represented in Fig. 5.47 [108].

A Heavy Fuel Oil (HFO) power plant dissipates approximately 30% of its total energy as waste heat in the surrounding, and the average monthly energy loss from a diesel engine is almost 17.30×10^5 MJ. Using this enormous wasted energy of a single diesel engine power plant, nearly 9.97×10^5 kg of food can be dried each month. Similarly, a massive amount of food can be dried using other sources of waste heat. Thus, a cleaner environment can be ensured by reducing the carbon footprint of food drying industries. The industrial aspect of waste heat is displayed in Fig. 5.48.

5.8.4.2   Performance Evaluation

Various performance parameters of the proposed waste heat drying system have been calculated using the data of the experiment performed by Masud et al. [206, 108, 109, 129]. The cost of the proposed system was estimated based on the design. Using the product value of equipment found in the local market, the price of a waste heat drying system was estimated to be around 154 USD [206]. Performance parameters such as cost, energy, entropy, emission, and payback period for potato drying were measured using a waste heat convective dryer. Summary of the results is presented in Fig. 5.49, which is drawn based on the experimental observations of Masud et al. [206]. However, in case of different sources and large-scale implementation the numerical values will vary significantly.

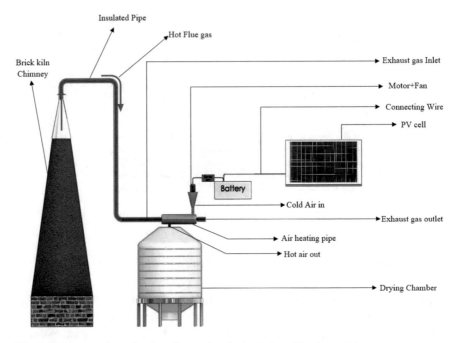

Fig. 5.47 Large-scale application of waste heat in the drying of food materials

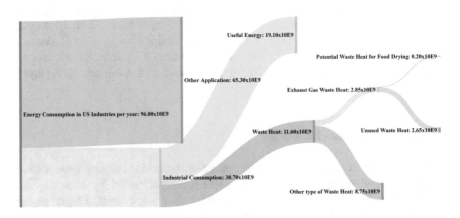

Fig. 5.48 Industrial aspect of recoverable waste heat

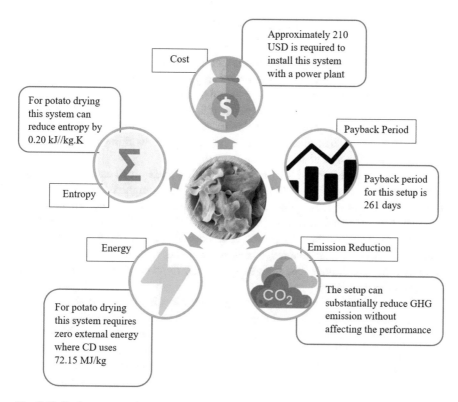

Fig. 5.49   Performance evaluation of a WHCD for potato drying

Almost 8.88 kg of CO_2 is produced when 3.78 liters of gasoline is burnt [130]. However, in the proposed technique, this substantial amount of GHG emission can be reduced without affecting the performance. From literature, it is evident that convective mushroom drying under 60 °C temperature with a total drying time of 184 minutes requires 2.5 kWh of energy [131]. Considering the fact that the system will return the cost of installment within 321 days, it can be considered as very cost-effective [206]. Therefore, this method of drying is highly recommended for low-income developing countries.

By implementation of the WHCD plant, heat carried out by waste heat was recovered, and some useful work was done in the process. By this process, the reduction of entropy is 0.20 kJ/kg.k, which was used to dry various samples [206]. Moreover, convective drying requires 14.53 MJ/kg energy for potato drying, but WHCD requires no external energy input at all. These results suggest that there is a significant prospect of WHCD in terms of reducing energy consumption and carbon footprint in the food processing industries [206].

5.9 Hybrid Solar-Biomass Drying

A hybrid solar-biomass dryer is essentially the combination of solar heating with biomass combustion as a backup thermal energy generator. The system is designed with transparent walls to ensure maximum solar irradiation allowance. Additionally, solar collectors can be used to improve the rate of drying [17, 132, 133]. Biomass heating system is used in off-sunshine hours. Since mixed-mode type solar dryer is considerably better than other types of solar drying, its working principle can be incorporated in the hybrid solar drying system.

Simate et al. investigated the performance of a mixed-mode natural convection solar dryer for maize drying [134]. It was concluded from the results that the mix mode dryer requires less collector area than conventional indirect mode dryers, and therefore, its installation cost is low. Besides, the mixed-mode dryer has more uniform moisture distribution. Forson et al. have also experimented using mixed-mode solar dryer for cassava chips drying [135]. They found that the drying rate of mixed-mode is higher than conventional solar dryers as it received direct and indirect solar irradiance.

Ideally, a hybrid solar-biomass drying system would employ solar irradiation as its primary source of thermal energy and use biomass as an auxiliary thermal source when solar radiation is unavailable. Therefore, in days with cloudy weather or during off-sunshine hours, the hybrid solar-biomass dryer can work splendidly. Thus, this drying mechanism can extend the drying period beyond sunshine hours. The mechanism of a hybrid solar-biomass is depicted using the flow chart in Fig. 5.50.

The system described in Fig. 5.50 gains its primary thermal energy from direct solar irradiation, which enters the dryer through the transparent cover. The solar radiation is absorbed by agricultural products and corrugated aluminium solar collector plate. As the collector plates heat up, it dissipates its thermal energy to the surrounding air that flows over the food products and absorbs moisture. Using chimney, a natural draft is maintained to ensure improved airflow. The moisture leaves the dryer through the chimney placed on top of the dryer. The natural draft created by this process depletes the pressure inside the dryer, which draws ambient air through the inlet section. When the solar radiation is not available, biomass is combusted, and its heat is directly supplied to the drying chamber. The improved temperature control system is expected for this type of dryer in order to ensure that food products are not being overheated.

5.10 Osmotic Dehydration

Organoleptic attributes such as elasticity, texture, juiciness, and tenderness are influenced by the amount of water present in food materials [136, 137]. Retaining these attributes after the drying process is essential for maintaining a good

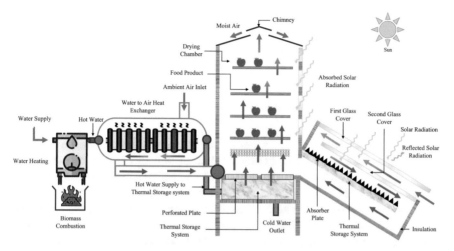

Fig. 5.50 Operating principle of hybrid solar-biomass drying

consumer perception. Additionally, it has been observed that the bound water of the food cells regulates the microbial growth and stability [138–141]. Exhaustive studies on food quality have revealed that the thermodynamic state influences the texture and stability as opposed to the quantity of water present in food. Researchers have coined the term water activity to define the thermodynamic state of water in food samples [142–144]. For completely dried food, the water activity is 0, and for pure water, the water activity is 1. The stability of food can be improved by lowering the water activity. Low water activity also changes the texture of food from elastic and juicy to crunchy and brittle.

Using the principle of osmosis, water can be removed from food samples in two ways: first, the process of solvent removal and second, humectants addition [145]. The second method is detrimental to the consumer's health since the process requires the addition of large amount of sugar, sodium chloride, or polyols with the food [146]. This method is also restrained by toxicological and nutritional concerns. Albeit the first process is expensive and energy-intensive, it provides an unparallel quality of dried food products. Therefore, the principle of solvent removal is employed for osmotic dehydration. Osmotic dehydration has some unique advantages. During the dehydration process, the quantity of Osmo-active substance that penetrates the food tissue can be regulated based on individual specification. The processed food samples do not require any rehydration for consumption. This process allows the freedom to regulate the food chemical composition according to requirements. The raw material mass can be reduced by half using this process.

Despite the numerous advantages, osmotic dehydration is incapable of a substantial reduction of microorganism proliferation, as it cannot deplete water activity sufficiently. Therefore, additional preservation methods like pasteurization, freezing, and drying are necessary to ensure adequate shelf life extension of food. Therefore, osmotic dehydration can be used as a pre-drying step, which has the potential to reduce the overall drying time and cost.

5.10.1 Osmotic Pressure

Interaction between solute and water characterizes the thermodynamic state of water. That is the energetic state of each molecule is influenced by its internal energy and interaction of surrounding particles. The energetic state of one-mole particle substance is defined as chemical potential. Therefore, the chemical potential is written as a function of temperature, pressure, and concentration. When the system operates isothermally, chemical potential only depends on pressure and concentration. Chemical active, which is decreased by the increase of solute concentration, can be calculated using the following equation [91]:

$$\mu_w = \mu_{ow} + RT \ln\left(a_w\right) \tag{5.5}$$

Where, μ_w = water chemical potential,
μ_{ow} = Standard state chemical potential
R = gas constant
T = absolute temperature
a_w = co-efficient of water activity

Energy exchange occurs when two systems with different levels of energy state interact with each other. The exchange of energy continues until both system's chemical potential reaches equilibrium.

5.10.2 Mass Transfer Is Osmotic Drying

A system is said to be in thermodynamic equilibrium when the cell and its surroundings have the same osmotic pressure. In this condition, the turgor pressure is also zero. Water is transferred out of a cell when the osmotic pressure of the cell's surrounding environment is higher than its inside, which causes the cell to shrink. This state is defined as a hypertonic solution. A cell subjected to a hypertonic solution will dissipate water. Through a process called plasmolysis, plasmalemma detaches from cell wall when protoplast is dehydrated (since it decreases its volume). The method of plasmolysis is revealed in Fig. 5.51. Hypertonic solution occupies the

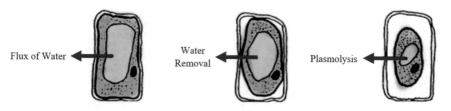

Fig. 5.51 Process of plasmolysis

space between the cell wall and plasmalemma due to the permeability of the cell wall. A layer of skin is removed from the food materials so that the cuticle and epidermal cells do not hinder the process of hypertonic osmosis.

5.10.3   Osmotic Dehydration Techniques for Food Products

Osmotic dehydration has considerable potential for energy saving since in this process water is removed from the food material without undergoing phase change. The amount of energy required to dry fruits and vegetables using this system is around 100–2400 kJ/kg of evaporated water [147]. This is significantly lower than convection drying, which requires 5 MJ/kg of evaporated water [91]. The high efficiency of osmotic dehydration makes it ideal for large scale operation. A snapshot of the feasible osmotic dehydration technologies for developing countries is presented here.

5.10.3.1   Osmotic Dehydration With Re-Concentration

Re-concentration system is essential to improve the efficiency of osmotic dehydration. As water is removed from the food material, it dilutes the hypertonic solution. Therefore, re-concentration is necessary either by Osmo-active substance dissolution or by excess water evaporation. Utilizing these methods of re-concentration, the hypertonic solution can be used repetitively. The solids and solution ratio influence the dilution of the hypertonic solution. When the ratio is about 1:10 (high), dilution is low, and vice versa [148]. It is possible to reuse osmotic drying hypertonic solution for around 20 times in a continuous process of osmotic dehydration. The recyclability depends on the type of processed materials, re-concentration technology, and parameters of pasteurization [148]. Schematic diagram of osmotic rehydration with re-concentration is represented in Fig. 5.52.

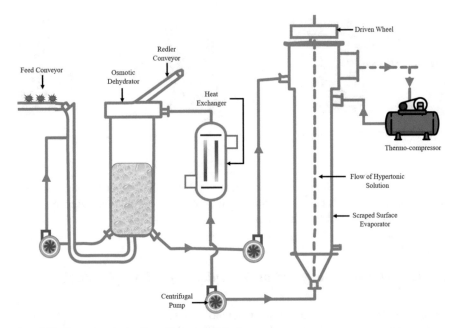

Fig. 5.52 Osmotic rehydration with re-concentration

5.10.3.2 Osmotic Dehydration With a Vibrating Plate Mixer

Decreases in mass transfer resistance can facilitate osmotic dewatering. This process can be utilized by slowly moving food or circulating the solution. Using a circulation pump and basket where food is immersed, solution circulation can be achieved. Food materials are moved through the solution using vibration. A schematic diagram of osmotic dehydration with a vibrating plate mixer is shown in Fig. 5.53.

The mechanism of vibrating plate mixer combines the principle of the slow movement of food and solution circulation. Using two loops, osmotic solution is circulated inside a vibrated plate mixer. One is called feed loop, and the other is used to control the temperature of the solution. Food is transported from the bottom of the mixer to its top.

5.11 Pulsed Microwave Solar Dryer

Researchers have been striving to improve energy efficiency and product quality in food drying for many years [149, 150]. Pulsed Microwave-Solar Dryer (PMSD), which is also known as Intermittent Microwave Convective Dryer (IMCD), is an approach that increases both energy efficiency and product quality [151–155].

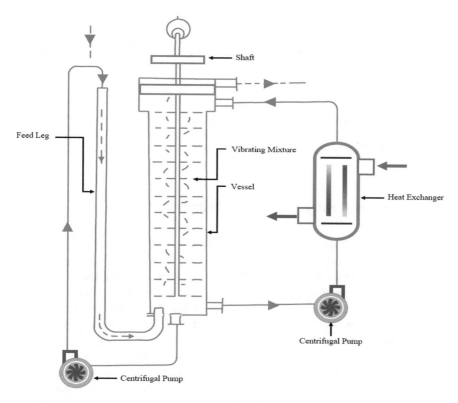

Fig. 5.53 A schematic diagram of osmotic dehydration with a vibrating plate mixture

Although there are some experimental investigations of PMSD, optimization of process parameters along with the incorporation of solar energy to run the system still remains under-developed. The optimization of PMSD is essential to develop an intelligent system that could manipulate the drying conditions (e.g., air temperature, flow rate, Relative Humidity, Microwave power, and Intermittency) in order to achieve higher energy efficiency and better quality [151, 156–158]. Therefore, a comprehensive PMSD setup to optimize the process of food preservation by utilizing renewable energy sources like solar energy can be a viable option to solve the issue.

A proper understanding of the changes in food material during processing is critical to choose the best drying conditions. However, the optimization of process parameters in PMSD has not been reported in the literature. Moreover, the maximum of the setups regarding PMSD that are designed and manufactured by the different researchers are too complicated. Therefore, to simplify the dryer, an intelligent control system is necessary that will control the process parameters for better quality retention and energy efficiency. The control system will automatically and instantaneously set the right drying condition based on the current quality and moisture level.

The proposed dryer, therefore, aims to develop an efficient and economically viable drying system capable of running independently of traditional energy sources. Moreover, the proposed PMSD system also offers fast and energy-efficient drying with high product quality.

The proposed PMSD setup has been constructed to overcome some of the limitations experienced by the researchers. The novelty of the proposed drying system can be found in its energy efficiency. Since the system is solar-assisted, it has low construction costs. Moreover, it is more accurate due to its microcontroller-based controller system. The system also has high degree of compactness and flexibility.

5.11.1 Basic Principle of Pulsed Microwave Drying

Microwaves are defined as electromagnetic radiation with 300 MHz–300GHz frequency range and with a wavelength range of 1 mm–1 m. Moisture inside food material can be heated up volumetrically when microwave penetrates it [159, 160]. Through this process, moisture content inside food material is removed as its diffusion rate and pressure gradient increases. Ionic conduction and bipolar re-orientation are the two primary mechanisms of microwave heating [161]. Water molecules can be vibrated with changing the electrical field due it its bipolar nature. Ideally, when 2.45 GHz microwave frequency is applied, the direction of the electric field is altered 2.45 billion times per second [162]. This extreme change in the electric field causes the water molecule to move erratically, which produces friction and generates a considerable amount of heat inside food [163]. Ionic conduction is a second major mechanism of microwave heating, which is caused by ions. Salty foods are most influenced by this phenomenon of the electric field.

By applying microwaves, the drying rate can be increased significantly, since it has a volumetric heating capacity. Advantages of microwave assisted drying are described below:

- **Volumetric heating**: Moisture diffusion rate increases and volumetric heating is accomplished when water molecules interact with microwave energy [164]. This can thereby significantly reduce drying times [163, 165]. It is clear from the literature that microwave convection (MWC) drying shows remarkably lower drying time than Convective Drying (CD) [164]–[166].
- **Controlled heating**: Since microwaves can be applied intermittently, it can control the fidelity of heating by changing the pulse ratio and the power level of the microwave [167] .
- **Selective heating**: Preferential heating of wetter areas is possible with microwave heating, and also bound water molecules can be excited by microwaves [164, 167–171].

5.11.1.1   Combined Microwave and Another Drying

Microwaves are combined with freeze-drying, osmotic drying, hot air-drying, and vacuum drying [172]–[173]. Due to less cost being involved, convective microwave drying (MCD) is accomplished by combining convection drying with a micro-wave [174].

Incorporation of microwaves (MW) with the CD can increase the supply of moisture to the outer surface of the food material, which will help to reduce the formation of crust and finally will assist in exacerbating the diffusion rate. This is the reason that MCD can significantly shorten drying time and improve product quality and energy efficiency [163]. When compared with convection drying, a sub-stantial reduction in the drying time (25–90%) has been found in MCD drying [175]–[176]. A number of studies found MCD treated products possess superior quality in contrast to conventional drying [175, 177]–[178].

5.11.1.2   Intermittent Application of MW in CD

Uneven heating and overheating of food materials are caused by improper power and energy supply. To overcome this problem, the introduction of MW energy inter-mittently can be an option. This intermittent application of MW maintains the expected temperature of the food materials during drying and the uniform distribu-tion of moisture within food materials, which improves product energy efficiency and quality [179]. Researchers have successfully combined vacuum drying with intermittent microwave energy [180]–[181] and freeze-drying [182], and they observed improved dried samples in comparison with vacuum and freeze-dried product. Additionally, it was found that PMSD is 4.7–11.2 times more energy-efficient and retains the food quality attributes such as color, texture, and smell bet-ter than other drying technologies [183].

Agricultural products such as pineapple [184], sage leaves [185]–[186], red pep-per [179], mushrooms [187], carrots, oregano [179] and bananas [188] can be dried using PMSD with splendid quality factor. Research conducted on the performance of PMSD is summarized in Table 5.7.

All the research that has been done on PMSD requires an external energy source to run. There is a lack of research in this field, although the pulsed microwave dry-ing treatments are less energy-intensive than the conventional dryer and require the lowest drying time than all the other traditional drying treatments. The introduction of renewable energy (e.g., solar energy) can be a substantial option to run the dryer in order to reduce conventional energy consumption. Thus, an innovative pulsed microwave solar-assisted convective dryer has been proposed to solve this issue.

Table 5.7 Summary of research regarding PMSD

Year	Material	System approach/ Methodology	Properties measured	Limitations	References
2006	Wheat and Rice	A bubbling type of thermo-hydrostatic dryer was used for drying. It consists of a heater, controller, drying tray, and humidifier. Temperature and humidity can be controlled, and the weight of the sample was measured every 10 min during every drying period.	1. Effect of temperature on tempering 2. Moisture content	1. Unable to measure the pressure difference 2. Not energy efficient 3. Incomplete control unit 4. Unable to measure the rehydration ratio.	[189]
2007	Banana	The experiment was carried out in the dryer, which consists of a microwave generator with 1000 w and 2450 MHz. Microwave power, air velocity, temperature, and humidity were an adjustable parameter.	1. Drying kinetics 2. Charring of plant foods	1. Not energy efficient 2. Unable to measure the pressure difference	[188]
2007	Agar gel	Microwave oven having two continuous output power settings of 250 W and 500 W operating at 2.45 GHz was used. Three microwave heating modes were studied; among them the first two were under the same average microwave power level with pulse ratio one and two, respectively. The third one is the same under the AP condition of 225 W.	1. Microwave power absorption 2. Temperature distribution	1. Unable to measure pressure differential 2. Not automated 3. Unable to measure moisture content 4. No well-designed control unit	[190]
2007	Mashed potato	Microwave oven having two continuous output power settings of 250 W and 500 W operating at 2.45 GHz was used.	1. Microwave penetration rate	1. Not energy efficient 2. Infirm control unit 3. Unable to measure the pressure difference 4. Unable to measure moisture content	[191]

(continued)

Table 5.7 (continued)

Year	Material	System approach/ Methodology	Properties measured	Limitations	References
2007	Carrots	A 750 W microwave unit operating at 2450 MHz provided with waveguides was used as a dryer. It consists of a strain gauge to measure real-time mass, a desiccator to remove moist air and optics fibers to measure internal temperature.	1. Quality analysis 2. Rehydration ratio	1. Unable to measure pressure differential 2. Not energy efficient	[187]
2009	Red pepper	Finding the effectiveness of various microwave–convective drying treatments of red pepper and comparing to convective air drying.	1. Drying kinetics 2. Physical properties (color and texture)	1. Power analysis 2. Temperature distribution	[183]
2009	Potato	The effectiveness of various microwave-convective air-drying treatments was compared to establish the most favorable drying condition of potato in terms of drying time, energy consumption & dried product quality.	1. Drying kinetics 2. Specific energy consumption 3. Sensory properties	1. No solar-assisted 2. No micro-controller unit	[192]
2010	Sage leaves	Intermittent drying was carried out with a dryer, including the speed of air at 1.2 m/s. the oven was set to the selected drying temperature for about 30 min to achieve the steady-state conditions before the samples in the oven.	1. Colour 2. Moisture content 3. Microwave distribution	1. Unable to measure pressure differential 2. Not energy efficient 3. Infirm control unit	[185]
2012	Pineapples	Temperature was measured using optical fibers of 4 pieces at different locations of the cavity. Different microwave power and pulsed ratio were used while conducting the experiment.	1. Temperature and microwave distribution 2. Product quality	1. Not energy efficient 2. Unable to measure pressure differential 3. Monitoring microstructure was not detailed. 4. Unable to measure the rehydration ratio.	[184]

(continued)

Table 5.7 (continued)

Year	Material	System approach/ Methodology	Properties measured	Limitations	References
2014	Apple	Development of an intermittent microwave heating model. Validation of the model done by comparing the temperature distribution. Investigated the temperature redistribution due to intermittency.	1. Temperature redistribution 2. Maximum temperature 3. Comparison between Experimental and simulation results	1. Theoretical modelling only 2. No micro-controller unit	[154]
2014	Carrots	Response surface methodology was used to observe the effect of microwave pulse ratio at the first drying stage, transition point of moisture content, and microwave pulse ratio of the second stage. Control system was provided to control the pulse ratio, temperature.	1. Colour 2. Determination of carotenes 3. Kinetics 4. Rehydration ratio	1. Unable to measure pressure differential 2. Not energy efficient	[193]
2015	Apple	The samples were dried in a conventional microwave oven for 60 seconds and then drying for 120 seconds in a convention dryer. Microwave oven supplied power levels of maximum 1100 W and 2.45 GHz frequency.	1. Equilibrium vapour pressure. 2. Temperature distribution 3. Moisture distribution 4. Absorbed power distribution	1. Not energy efficient 2. No micro-controller unit 3. Unable to measure rehydration ratio	[153]
2015	Apple	A domestic microwave oven operating at 1100 W nominal power with 2465 Mhz was used as a dryer with four heating units and a centrifugal fan. The body was thermally insulated. A cycle controller was also used for controlling the magnetron.	1. Rehydration ratio 2. Phenolic content 3. Color 4. Energy consumption 5. Bulk density 6. The response of surface analysis	1. Pressure differential 2. Monitoring microstructure 3. Not energy efficient	[194]

(continued)

Table 5.7 (continued)

Year	Material	System approach/ Methodology	Properties measured	Limitations	References
2016	Pumpkin	The drying of samples by microwave, convective, and intermittent microwave–convective techniques was studied. The samples were pre-treated by blanching, followed by pulsed vacuum osmotic dehydration. The microwave output power was 780 W.	1. Experimental behavior 2. Modelling of drying kinetics 3. Quality analysis	1. Incomplete control unit 2. Not solar assisted 3. No TIC & IC 4. Not innovative	[195]
2016	Lemon slice	Drying behavior of lemon slices was investigated using convective (50, 55 and 60 °C inlet air temperatures), microwave (specific power of 0.97 Wg-^{1}), and combined microwave-convective (specific powers of 0.97 and 2.04 Wg-^{1} assisted with 50, 55 and 60 °C inlet air temperatures) dehydration methods.	1. Drying kinetics 2. Determination of rehydration capacity 3. Modelling of drying behavior	1. Theoretical 2. Not energy efficient 3. No temperature distribution	[196]
2016	Granny smith apples	A conventional microwave oven operating at 1100 W and 2.45 GHz frequency was used as a dryer. Food was dried for 20 s in the microwave oven after 80 s in convection dryer.	1. Temperature distribution 2. Moisture distribution 3. Pressure distribution	1. Poorly developed control unit 2. Not able to monitor the microstructure 3. Not energy efficient 4. Unable to measure bound water.	[155]
2017	Kiwi fruit	IMCD drying was carried out for apple slices by using a microwave oven for the 20 s then 80 s in the convention dryer.	1. Temperature distribution 2. Moisture distribution 3. Pressure distribution	1. Not energy efficient 2. Infirm control unit	[197]

5.11.2   Experimental Design and Fabrication

A Pulsed Microwave Solar Drying system has been designed and fabricated by Joardder et al., which has been presented in Fig. 5.54 [198]. Domestic microwave oven (1050 W) is used as a source of Microwave. The system was amended to have a nominal microwave (700 W). It is controlled using a cross-sectioned connective dryer of 200 W power. To control the velocity of the convective dryer, there is a regulator attached to the setup.

In order to supply hot air to the chamber, an opening has been made on the left wall of the dryer. The moist air is removed with the vents in the right wall of the system. Using an automated control system, the intermittency of microwaves in the drying system is maintained in the desired time interval. The entire setup can be placed in a Table ($4m \times 2.5m \times 3m$) that can achieve the compactness of the system. The entire power supply has been carried out by using a solar panel as a power source to minimize the supply of the external powers. A battery is used which will be charged with solar panel and inverter converts the DC into AC.

5.11.2.1   Design Criteria

PMSD was designed based on some specified criteria to ensure maximum performance and economic feasibility. A comprehensive discussion of the design criteria of PMSD is presented in this section. The summary of the design criteria for the PMSD setup has been extensively represented in Fig. 5.55.

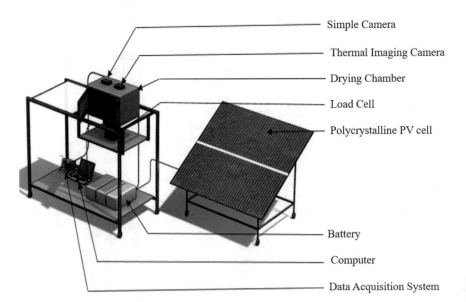

Fig. 5.54 Experimental setup of PMSD (PV cell is used to place in the rooftop. The photo-taking purpose it is placed beside the setup)

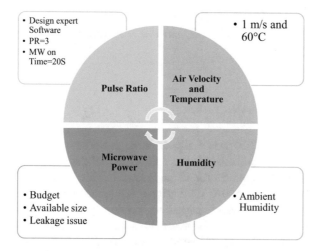

Fig. 5.55 Summary of the design criteria for PMSD set-up

Air Velocity and Temperature Different renowned authors have done extensive experiments to optimize the proper drying conditions [199–203]. For the proposed PMSD, the best-optimized result has been selected to design the PMSD. Tzempelikos et al. have found that when the temperature is increased from 40 °C to 60 °C, the total drying time is decreased by 54%. Similarly, at 60 ° C, when the velocity is increased from 1 ms^{-1} to 2 ms^{-1}, the total drying time is decreased by 30% [204]. Therefore, in this study, to design the PMSD air temperature and velocity was selected 60 °C and 1 ms^{-1}, respectively.

Humidity Relative humidity of the air is another critical factor for drying. For the designed PMSD, the relative humidity was not manipulated. The standard ambient humidity was selected for the subsequent experiments.

Microwave Power Design and optimization of microwave power and size are the two biggest issues in designing the setup. There is basically some basis on which the microwave power was designed for the proposed PMSD

- **Available Size:** Designing a new Microwave is very critical due to the unavailability of material. Therefore, the best suited available size was selected that may be able to fulfill all the requirements.
- **Leakage Problem:** Microwave leakage is one of the main issues in designing the Microwave. As it is almost an unsolvable challenge; therefore, the available designed Microwave was selected to accomplish subsequent experiments.

Pulse Ratio Pulse ratio optimization is one of the main factors in PMSD. Although there is some literature on PMSD and pulse ratio 2, 3, and 4 can be formed in several ways. There is a countless set of times by which the pulse ratio can be selected.

Therefore, by using the Design Expert software, the set of time that is suitable for the fabricated set-up was selected, and that was verified by doing an extensive amount of experimental study.

5.11.3 Feasibility Analysis

Performance of the PMSD was evaluated through rigorous experimentations. The cost of the dryer was estimated based on the design of the proposed system. Using the product value of equipment found the in local market, the price of a PMSD system was estimated to be 600 USD. Performance parameters such as cost, energy, entropy, emission, and payback period for potato drying were measured using a PMSD. Summary of the results is presented in Fig. 5.56, which is drawn based on fabricated experimental setup of Joardder et al. [207]. However, in case of different sources and large scale implementation the numerical values will vary significantly.

The proposed PMSD has the potential to reduce 510 tons of CO_2 emission annually if apple is to be dried. This will considerably reduce the carbon footprint for

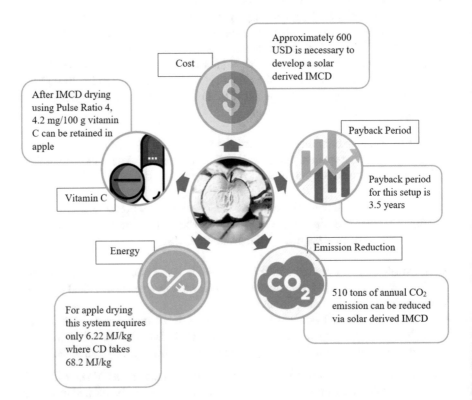

Fig. 5.56 Performance evaluation of PMSD for apple drying

apple drying and mediate the effects of global warming. Considering the fact that the system will return the cost of installment within 3.5 years, it can be considered as cost-effective. Therefore, this method of drying is highly recommended for middle-income developing countries.

This system can significantly retain the vitamin content after a specific food material has been dried. The data found in the case of apple drying shows that, after PMSD drying, apple samples retain 4.1 ml/100 g Vitamin C (almost 91% of fresh apple's vitamin C). It retains nearly 45.7% more Vitamin C than convective drying. Moreover, PMSD requires only 6.22 MJ/kg for apple drying; in contrast to this, convective drying (CD) takes a staggering 50 MJ/kg of energy. These results suggest that there is a significant prospect of PMSD in terms of reducing energy consumption and carbon footprint in the food processing industries.

5.12 Drying Technique Selection

In the food processing industry, most food products experience drying at one or more stages of its processing. Moisture content of the product also affects the cost of transport (in case of coal), and a balance between the cost of drying, and the cost of conveying should be maintained. However, over-drying causes degradation of product quality (in case of paper and timber), wastes energy by producing more heat and is highly expensive. So, the methods which are used to save energy in dryers should be taken into consideration. This study highlights the paradigm of products and precautions types of drying.

In the case of manufacturing colors and dyes, because of their heat sensitivity, thermal drying is a necessary step. For having proper shade of color, and the minimization of thermal degradation, the effect of time and temperature in drying must be in focus. Lower temperature and nonexistence of air can make drying successful. Dryer of recirculation type truck and tray compartment is mostly used worldwide. In pharmaceutical and chemical industries, drying is typically required before packaging. Therefore, different products require different types of dryers; products such as antibiotics and blood plasma, which are highly heat-sensitive, need high vacuum tray drying or freeze-drying. Similarly, large tonnages of minerals, heavy chemicals, and natural ores need continuous rotary dryers, through circulation and large turbo-tray dryers are also used. Compatibility of various types of food with their respective drying techniques is displayed in Table 5.8. Using the following table, dryers can be selected based on the type of food material.

As mentioned above, the end moisture content is largely important in drying as it determines the stability and storage requirement of product. It also dictates the conditions as well as drying time. The variations between interparticle moisture content and internal moisture gradients are important. Moisture content also is linearly related to the extent of shrinkage of many hygroscopic materials (which may shrink in drying), and it can change the hygroscopic limit. Drying time is an essential factor while selecting a dryer as the cost of drying is directly associated with it. Using Table 5.9, drying techniques can be selected based on the desired time.

Table 5.8 Selection of dryer based on the type of food materials

Type of food material	Solids (Formed)	Solids (Free Flowing)					Cakes		Liquids		
		Fiber	Fragile Crystal	Pellet	Granule	Powder	Filter	Centrifuge	Solution	Paste	Slurry
Improved Solar Drying											
Geothermal Drying											
Hybrid Geothermal Drying											
Heat Pump Drying											
Drying Using Waste heat											
Desiccant Drying											
Improved Biomass Drying											
Hybrid Solar-Biomass Drying											
Solar Derived IMCD											
Osmotic Drying											

Legend: Recommended   Conditionally Recommended   Not Recommended

Temperature causes phase changes, degradation, flammability of dust, discoloration, staining, and many other problems. However, if maximum temperature is fixed by thermal sensitivity, it will be useful for the substance to be exposed at that temperature during the drying time. In some dryers like pneumatic and spray drying, heat-sensitive substances are permitted during the preservation time at higher temperatures. Some problems arise in tensile cracking and case hardening. There are available details on defeating these problems by compromising between drying time and energy efficiency.

The proposed dryers discussed in Chap. 5 are designed and developed in order to improve the drying efficiency and reduce the cost of post food production. These proposed new technologies developed based on the authors' extensive research, experimental investigation, and computing various performance parameters of these dryers. Despite many alluring advantages, when adopting these technologies, authorities might face numerous challenges. In order to prepare them for said challenges, these challenges for each of the proposed dryers are discussed in Chap. 6. The constant development of science, technology, and metallurgy may one day solve all those challenges.

Acknowledgement Icons used in the illustration of various figures of this chapter were made by Freepik, Itim2101, Picol, Smashicons, Monkin, Phatplus, Vector Market, and Good Ware from www.flaticon.com.

Table 5.9 Categorization of dryer based on cost and drying time

Type of Dryer	Drying Time					Cost of Installation		
	0–30 (seconds)	5–10 (minutes)	30–60 (minutes)	1–6 (hours)	10+ (hours)	Low	Medium	High
Improved Solar Drying					■	■		
Geothermal Drying					■			■
Hybrid Geothermal Drying			■					■
Heat Pump Drying		■	■				■	
Drying Using Waste heat			■			■		
Desiccant Drying					■	■		
Improved Biomass Drying	■	■					■	■
Hybrid Solar-biomass Drying	■		■			■	■	
Solar Derived IMCD	■							■
Osmotic Drying				■		■		

References

1. M.U.H. Joardder, M.H. Masud, *Food Preservation in Developing Countries: Challenges and Solutions,* Springer (2019)
2. M.U.H. Joardder, M.H. Masud, Challenges and mistakes in food preservation, in *Food Preservation in Developing Countries: Challenges and Solutions*, Springer, 175–198 (2019)
3. M.U.H. Joardder, M.H. Masud, Possible solution of food preservation techniques, in *Food Preservation in Developing Countries: Challenges and Solutions*, Springer, 199–218 (2019)
4. M.M. Rahman, M.U.H. Joardder, M.I.H. Khan, N.D. Pham, M.A. Karim, Multi-scale model of food drying: Current status and challenges. Crit. Rev. Food Sci. Nutr. **58**(5), 858–876 (2018)
5. M.M. Rahman, Y.T. Gu, M.A. Karim, Development of realistic food microstructure considering the structural heterogeneity of cells and intercellular space. Food Struct. **15**, 9–16 (2018)

6. M.M. Rahman, M.U.H. Joardder, A. Karim, Non-destructive investigation of cellular level moisture distribution and morphological changes during drying of a plant-based food material. Biosyst. Eng. **169**, 126–138 (2018)
7. M.U.H. Joardder, M.A. Karim, Development of a porosity prediction model based on shrinkage velocity and glass transition temperature. Dry. Technol., 1–17 (2019)
8. M.U.H. Joardder, C. Kumar, M.A. Karim, Prediction of porosity of food materials during drying: Current challenges and directions. Crit. Rev. Food Sci. Nutr. **58**(17), 2896–2907 (2018)
9. A.S. Mujumdar, A.S. Menon, Drying of solids: Principles, classification, and selection of dryers. Handb. Ind. Dry. **1**, 1–39 (1995)
10. A.S. Mujumdar, Classification and selection of industrial dryers. Mujumdar's Pract. Guid. to Ind. Dry. Princ. Equip. New Dev. Bross. Canada Exergex Corp., 23–36 (2000)
11. S.A. Klein, Calculation of flat-plate collector loss coefficients. Sol. Energy **17**, 79 (1975)
12. M.A. Basunia, T. Abe, Thin-layer solar drying characteristics of rough rice under natural convection. J. Food Eng. **47**(4), 295–301 (2001)
13. P. Barnwal, G.N. Tiwari, Grape drying by using hybrid photovoltaic-thermal (PV/T) greenhouse dryer: An experimental study. Sol. Energy **82**(12), 1131–1144 (2008)
14. B.M.A. Amer, M.A. Hossain, K. Gottschalk, Design and performance evaluation of a new hybrid solar dryer for banana. Energy Convers. Manag. **51**(4), 813–820 (2010)
15. S.C. Bhattacharya, T. Ruangrungchaikul, and H.L. Pham, Chapter 240 – Design and Performance of a Hybrid Solar/Biomass Energy Powered Dryer for Fruits and Vegetables, A.A.M.B.T.-W.R. E.C. V.I. Sayigh, Ed. Oxford: Pergamon, 1161–1164 (2000)
16. M.A. Karim, E. Perez, Z.M. Amin, Mathematical modelling of counter flow v-grove solar air collector. Renew. Energy **67**, 192–201 (2014)
17. M.A. Karim, M.N.A. Hawlader, Performance investigation of flat plate, v-corrugated and finned air collectors. Energy **31**(4), 452–470 (2006)
18. M. Islam, S. Miller, P. Yarlagadda, A. Karim, Investigation of the effect of physical and optical factors on the optical performance of a parabolic trough collector. Energies **10**(11), 1907 (2017)
19. M. Islam, M.A. Karim, S.C. Saha, S. Miller, P.K.D.V. Yarlagadda, Development of empirical equations for irradiance profile of a standard parabolic trough collector using Monte Carlo ray-tracing technique, in *Advanced Materials Research*, **860**,180–190 (2014)
20. M. Islam, P. Yarlagadda, A. Karim, Effect of the orientation schemes of the energy collection element on the optical performance of a parabolic trough concentrating collector. Energies **12**(1), 128 (2019)
21. D. Heim, P. Klemm, Numerical solution of TIM-PCM solar thermal storage system with ESP-r. Res. Build. Phys., 683–690 (2003)
22. K.S. Ong, Thermal performance of solar air heaters: Mathematical model and solution procedure. Sol. Energy **55**(2), 93–109 (1995)
23. F.K. Forson, M.A.A. Nazha, H. Rajakaruna, Experimental and simulation studies on a single pass, double duct solar air heater. Energy Convers. Manag. **44**(8), 1209–1227 (2003)
24. M.U.H. Joardder, M.H. Masud, Feasibility of advance technologies, in *Food Preservation in Developing Countries: Challenges and Solutions*, Springer, 219–236 (2019)
25. L.M. Bal, S. Satya, S.N. Naik, Solar dryer with thermal energy storage systems for drying agricultural food products: A review. Renew. Sust. Energ. Rev. **14**(8), 2298–2314 (2010)
26. M.A. Karim, Experimental investigation of a stratified chilled-water thermal storage system. Appl. Therm. Eng. **31**(11–12), 1853–1860 (2011)
27. M.A. Karim, O. Arthur, P.K.D.V. Yarlagadda, M. Islam, M. Mahiuddin, Performance investigation of high temperature application of molten solar salt nanofluid in a direct absorption solar collector. Molecules **24**(2), 285 (2019)
28. A. Karim, A. Burnett, S. Fawzia, Investigation of stratified thermal storage tank performance for heating and cooling applications. Energies **11**(5), 1049 (2018)
29. O. Arthur, M.A. Karim, An investigation into the thermophysical and rheological properties of nanofluids for solar thermal applications. Renew. Sust. Energ. Rev. **55**, 739–755 (2016)

30. A.J. Mahmood, L.B.Y. Aldabbagh, F. Egelioglu, Investigation of single and double pass solar air heater with transverse fins and a package wire mesh layer. Energy Convers. Manag. **89**, 599–607 (2015)

31. P. Naphon, On the performance and entropy generation of the double-pass solar air heater with longitudinal fins. Renew. Energy **30**(9), 1345–1357 (2005)

32. H. Yeh, T.-T. Lin, Efficiency improvement of flat-plate solar air heaters. Energy **21**(6), 435–443 (1996)

33. A.E. Kabeel, A. Khalil, S.M. Shalaby, M.E. Zayed, Experimental investigation of thermal performance of flat and v-corrugated plate solar air heaters with and without PCM as thermal energy storage. Energy Convers. Manag. **113**, 264–272 (2016)

34. J.M. Troeger, J.L. Butler, Simulation of solar peanut drying. Trans. ASAE **22**(4), 906–911 (1979)

35. H.P. Garg, V.K. Sharma, R.B. Mahajan, A.K. Bhargave, Experimental study of an inexpensive solar collector cum storage system for agricultural uses. Sol. Energy **35**(4), 321–331 (1985)

36. S.O. Enibe, Performance of a natural circulation solar air heating system with phase change material energy storage. Renew. Energy **27**(1), 69–86 (2002)

37. M.M. Farid, A.M. Khudhair, S.A.K. Razack, S. Al-Hallaj, A review on phase change energy storage: Materials and applications. Energy Convers. Manag. **45**(9), 1597–1615 (2004)

38. V.V. Tyagi, N.L. Panwar, N.A. Rahim, R. Kothari, Review on solar air heating system with and without thermal energy storage system. Renew. Sust. Energ. Rev. **16**(4), 2289–2303 (2012)

39. A. Sharma, V.V. Tyagi, C.R. Chen, D. Buddhi, Review on thermal energy storage with phase change materials and applications. Renew. Sust. Energ. Rev. **13**(2), 318–345 (2009)

40. H.E.S. Fath, Thermal performance of a simple design solar air heater with built-in thermal energy storage system. Energy Convers. Manag. **36**(10), 989–997 (1995)

41. E.-B.S. Mettawee, G.M.R. Assassa, Experimental study of a compact PCM solar collector. Energy **31**(14), 2958–2968 (2006)

42. V.V. Tyagi, A.K. Pandey, S.C. Kaushik, S.K. Tyagi, Thermal performance evaluation of a solar air heater with and without thermal energy storage. J. Therm. Anal. Calorim. **107**(3), 1345–1352 (2012)

43. M.M. Alkilani, K. Sopian, S. Mat, M.A. Alghoul, Output air temperature prediction in a solar air heater integrated with phase change material. Eur. J. Sci. Res. **27**(3), 334–341 (2009)

44. A.A. El-Sebaii, S. Aboul-Enein, M.R.I. Ramadan, S.M. Shalaby, B.M. Moharram, Investigation of thermal performance of-double pass-flat and v-corrugated plate solar air heaters. Energy **36**(2), 1076–1086 (2011)

45. A.A. El-Sebaii, S. Aboul-Enein, M.R.I. Ramadan, S.M. Shalaby, B.M. Moharram, Thermal performance investigation of double pass-finned plate solar air heater. Appl. Energy **88**(5), 1727–1739 (2011)

46. A. Ayensu, V. Asiedu-Bondzie, Solar drying with convective self-flow and energy storage. Sol. Wind Technol. **3**(4), 273–279 (1986)

47. G.N. Tiwari, A.K. Singh, P.S. Bhatia, Experimental simulation of a grain drying system. Energy Convers. Manag. **35**(5), 453–458 (1994)

48. P.M. Chauhan, C. Choudhury, H.P. Garg, Comparative performance of coriander dryer coupled to solar air heater and solar air-heater-cum-rockbed storage. Appl. Therm. Eng. **16**(6), 475–486 (1996)

49. T. Ziegler, I.-G. Richter, R. Pecenka, Desiccant grain applied to the storage of solar drying potential. Dry. Technol. **17**(7–8), 1411–1427 (1999)

50. D.B. Jani, M. Mishra, P.K. Sahoo, Solid desiccant air conditioning–a state of the art review. Renew. Sust. Energ. Rev. **60**, 1451–1469 (2016)

51. D. Jain, Modeling the performance of the reversed absorber with packed bed thermal storage natural convection solar crop dryer. J. Food Eng. **78**(2), 637–647 (2007)

52. A. Madhlopa, G. Ngwalo, Solar dryer with thermal storage and biomass-backup heater. Sol. Energy **81**(4), 449–462 (2007)

53. P. Muffler, R. Cataldi, Methods for regional assessment of geothermal resources. Geothermics **7**(2–4), 53–89 (1978)

54. P.F.A. Ogola, The Power to Change: Creating Lifeline and Mitigation-Adaptation Opportunities through Geothermal Energy Utilisation, University of Iceland, (2013)
55. N.C. Vasquez, R.O. Bernardo, R.L. Cornelio, Industrial uses of geothermal energy a framework for application in a developing country. Geothermics **21**(5–6), 733–743 (1992)
56. S. Arason, The drying of fish and utilization of geothermal energy; the Icelandic experience. In International Geothermal Conference, Reykjavík, September (2003)
57. K. Popovski, K. Dimitrov, B. Andrejevski, S. Popovska, Geothermal rice drying unit in Kotchany, Macedonia. Geothermics **21**(5–6), 709–716 (1992)
58. N. Andritsos, P. Dalampakis, N. Kolios, Use of geothermal energy for tomato drying. GHC Bull. **24**(1) (2003)
59. M. Van Nguyen, S. Arason, M. Gissurarson, P. G. Pálsson, Uses of geothermal energy in food and agriculture opportunities for developing countries, Food And Agriculture Organization of The United Nations, Rome, (2015)
60. S. Popovska, D.S. Jayas, C.B. Singh, Drying of agricultural products with geothermal energy. Encycl. Earth Sci. Ser. **Part 4**, 231–232 (2011)
61. J.W. Lund, M.A. Rangel, Pilot fruit drier for the los Azufres geothermal field, Mexico, in *Processing of the World Geothermal Congress*, Florence, Italy, 18–31 (1995)
62. P. Mangi, Geothermal Resource Optimization: A Case of the Geothermal Health Spa and Demonstration Centre at the Olkaria Geothermal Project, Presented at Short Course VII on Exploration for Geothermal Resources, organized by UNU-GTP, GDC and KenGen, at Lake Bogoria and Lake Naivasha, Kenya, Oct. 27 – Nov. 18, (2012)
63. K. Abdullah, I.B.P Gunadnya, Use of geothermal energy for drying and cooling purposes. In Proceedings World Geothermal Congress, 1–5, April (2010)
64. J.W. Lund, Direct heat utilization of geothermal resources worldwide 2005. ASEG Ext. Abstr. **2006**(1), 1–15 (2006)
65. S. Björnsson, *Geothermal Development and Research in Iceland, By National Energy Authority and Ministries of Industry and Commerce* (2006)
66. J.W. Lund, D.H. Freeston, T.L. Boyd, Direct Utilization of Geothermal Energy 2010 Worldwide Review. Proceedings of the World Geothermal Congress. *Horne (ed.)*, Int. Geotherm. Assoc. Nusa Dua, Bali, **25**(4) (2010)
67. M.S. Uddin, M.H. Masud, S. Mandal, M. Morshed, Construction and performance study of underground assisted air heating and cooling system, in *International Conference on Mechanical, Industrial and Materials Engineering*, RUET, Rajshai, Bangladesh (2015)
68. L.F. Cabeza, A. Castell, C. Barreneche, A. De Gracia, A.I. Fernández, Materials used as PCM in thermal energy storage in buildings : A review. Renew. Sust. Energ. Rev. **15**(3), 1675–1695 (2011)
69. Y. Zhang, Research of Thermal Energy Storage Technology in the Solar Thermodynamic Power. Journal of Power and Energy Engineering, **4**(07), 42 (2016)
70. M.M. Farid, A.M. Khudhair, S.A.K. Razack, S. Al-Hallaj, A review on phase change energy storage: materials and applications. Energy conversion and management, **45**(9–10), 1597–1615 (2004)
71. P. Suntivarakorn, S. Satmarong, C. Benjapiyaporn, S. Theerakulpisut, An experimental study on clothes drying using waste heat from split type air conditioner. International Journal of Mechanical, Aerospace, Industrial, Mechatronic and Manufacturing Engineering, **3**(5), 483–488 (2009)
72. R. Tuğrul Oğulata, Utilization of waste-heat recovery in textile drying. Appl. Energy **79**(1), 41–49 (2004)
73. H. Li, Q. Chen, X. Zhang, K.N. Finney, V.N. Sharifi, J. Swithenbank, Evaluation of a biomass drying process using waste heat from process industries: A case study. Appl. Therm. Eng. **35**, 71–80 (2012)
74. Y. Qin, H. Fu, J. Wang, M. Liu, J. Yan, Waste heat and water recovery characteristics of heat exchangers for dryer exhaust. Dry. Technol. **36**(6), 709–722 (2018)
75. T.M.I. Mahlia, L.W. Cheng, L.C.S. Salikka, C.L. Lim, M.H. Hasan, U. Hamdani, Drying garcinia atroviridis using waste heat from condenser of a split room air conditioner. Int. J. Mech. Mater. Eng. **7**(2), 171–176 (2012)

76. J.W. MacArthur, E.W. Grald, Unsteady compressible two-phase flow model for predicting cyclic heat pump performance and a comparison with experimental data. Int. J. Refrig. **12**(1), 29–41 (1989)

77. J.W. MacArthur, Transient heat pump behaviour: A theoretical investigation. Int. J. Refrig. **7**(2), 123–132 (1984)

78. M. Alves-Filho, I. Stranmen, The application of heat pump in drying of biomaterials. Dry. Technol. **14**(9), 2061–2090 (1996)

79. S.J. Rossi, L.C. Neues, T.G. Kicokbusch, Thermodynamic and energetic evaluation of a heat pump applied to the drying of vegetables. Drying **92**, 475–478 (1992)

80. A.A. Nassikas, C.B. Akritidis, A.S. Mujumdar, Close-cycle heat pump dryer using super heated steam: An application to paper drying. Drying **92**, 1085–1098 (1992)

81. J.P. Meyer, G.P. Greyvenstein, The drying of grain with heat pumps in South Africa: A techno-economic analysis. Int. J. Energy Res. **16**(1), 13–20 (1992)

82. I. Strommen, K. Joseffsen, K. Kramer, Heat Pump Fluidised Bed Drying of Biologically Active Solutions, in *Drying'94—Proceedings of the 9th International Drying Symposium*, 1007–1014 (1994)

83. R.L. Mason, A.V. Blarcom, Drying macadamia nuts using a heat pump dehumidifier, in *The Development and Application of Heat Pump Dryers. Seminar Papers*, 1–7, 24th March, (1993)

84. C.G. Carrington, P. Bannister, An empirical model for a heat pump dehumidifier drier. Int. J. Energy Res. **20**(10), 853–869 (1996)

85. S. Prasertsan, P. Saen-Saby, Heat pump drying of agricultural materials. Dry. Technol. **16**(1–2), 235–250 (1998)

86. S.K. Chou, K.J. Chua, New hybrid drying technologies for heat sensitive foodstuffs. Trends Food Sci. Technol. **12**(10), 359–369 (2001)

87. U.S. Pal, M.K. Khan, S.N. Mohanty, Heat pump drying of green sweet pepper. Dry. Technol. **26**(12), 1584–1590 (2008)

88. S. Şevik, M. Aktaş, H. Doğan, S. Koçak, Mushroom drying with solar assisted heat pump system. Energy Convers. Manag. **72**, 171–178 (2013)

89. C.L. Hii, C.L. Law, S. Suzannah, Drying kinetics of the individual layer of cocoa beans during heat pump drying. J. Food Eng. **108**(2), 276–282 (2012)

90. Q. Shi, Y. Zheng, Y. Zhao, Mathematical modeling on thin-layer heat pump drying of yacon (Smallanthus sonchifolius) slices. Energy Convers. Manag. **71**, 208–216 (2013)

91. A.S. Mujumdar, *Handbook of Industrial Drying*, CRC press (2014)

92. B. Geeraert, Nato advance study institute series, series E. Appl. Sci. **1**(15), 219 (1976)

93. C. Strumillo, R. Zylla, *Drying'85, Mujumdar, AS*, Elsevier Science, Amsterdam (1985)

94. D.J. Barr, C.G.J. Baker, Specialized drying systems (Vol. 196). Chapman & Hall: New York, (1997)

95. S. Prasertsan, P. Saen-Saby, Heat pump dryers: Research and development needs and opportunities. Dry. Technol. **16**(1–2), 251–270 (1998)

96. K.J. Chua, A.S. Mujundar, S.K. Chou, M.N.A. Hawlader, J.C. Ho, Heat pump drying of banana, guava and potatoes pieces: Effect of cyclical variations of air temperature on convective drying kinetics on color changes. Dry. Technol. **18**, 907–936 (2000)

97. O. Alves-Filho, I. Strommen, Performance and improvements in heat pump dryers. Drying **96**, 405–415 (1996)

98. A.S. Mujumdar, S. Devahastin, *Developments in Drying: Food Dehydration*, vol 1, Kasetsart University Press, Bangkok, Thailand (1999)

99. A. Barati, M. Kokabi, M.H.N. Famili, Drying of gelcast ceramic parts via the liquid desiccant method. J. Eur. Ceram. Soc. **23**(13), 2265–2272 (2003)

100. J. W. Twidell, J. Muniba, T. Thornwa, The Strathclyde solar crop dryer: air heater, photovoltaic fan and desiccants, in *Proceeding of ISES solar world congress*, vol. 8, 55–60, (1993)

101. K. Nagaya, Y. Li, Z. Jin, M. Fukumuro, Low-temperature desiccant-based food drying system with airflow and temperature control. J. Food Eng. **75**, 71–77 (2006)

102. K.J. Chua, S.K. Chou, Low-cost drying methods for developing countries. Trends Food Sci. Technol. **14**(12), 519–528 (2003)

103. M.H. Masud, A.A. Ananno, P. Dabnichki, N. Ahmed, M. Mahjabeen, Prospect of Chicken litter as a source of sustainable energy, in *Technologies for Manure Conversion and Recycling,* Springer, in press (2020)
104. M.H. Masud, A.A. Ananno, P. Dabnichki, S. Hossain, S.A. Chowdhury, Anaerobic Co-digestion of Food Waste with Liquid Dairy Manure, in *Technologies for Manure Conversion and Recycling,* Springer, in press, (2020)
105. M.H. Masud, A.A. Ananno, A.M.E. Arefin, R. Ahamed, P. Das, M.U.H. Joardder, Perspective of biomass energy conversion in Bangladesh. Clean Techn. Environ. Policy **21**(4) (2019)
106. C M. Van't Land, Drying in the process industry. John Wiley & Sons. (2012)
107. M.U.H. Joardder, M.H. Masud, Food preservation techniques in developing countries, in *Food Preservation in Developing Countries: Challenges and Solutions*, Springer, 67–125 (2019)
108. M.H. Masud, T. Islam, M.U.H. Joardder, A.A. Ananno, P. Dabnichki, CFD analysis of a tube-in-tube heat exchanger to recover waste heat for food drying. Int. J. Energy Water Resour., 1–18 (2019)
109. M.H. Masud, M.T. Islam, A.A. Ananno, M.A. Ahmed, Towards a zero energy based food drying system by utilizing the waste heat, in *International Conference on Engineering Research, Innovation and Education*, Sylet, Bangladesh (2019)
110. A.M.E. Arefin, M.H. Masud, M.U.H. Joardder, M. Mourshed, F.R. Naim-Ul-Hasan, Waste heat recovery systems for internal combustion engines: A review, in *1st International Conference on Mechanical Engineering and Applied Science, At Military Institute of Science and Technology*, Dhaka, 1–4 (2017)
111. M. Karvonen, R. Kapoor, A. Uusitalo, V. Ojanen, Technology competition in the internal combustion engine waste heat recovery: A patent landscape analysis. J. Clean. Prod. **112**, 3735–3743 (2016)
112. K. Tuck, Researcher aims to use waste heat to make cars more efficient, *Boise State University* (2013)
113. J.S. Jadhao, D.G. Thombare, Review on exhaust gas heat recovery for I.C. engine. Certif. Int. J. Eng. Innov. Technol. **9001**(12), 2277–3754 (2008)
114. B.-T. Liu, K.-H. Chien, C.-C. Wang, Effect of working fluids on organic Rankine cycle for waste heat recovery. Energy **29**(8), 1207–1217 (2004)
115. R. Saidur, M. Rezaei, W.K. Muzammil, M.H. Hassan, S. Paria, M. Hasanuzzaman, Technologies to recover exhaust heat from internal combustion engines. Renew. Sust. Energ. Rev. **16**(8), 5649–5659 (2012)
116. A.V. Mehta, R.K. Gohil, J.P. Bavarva, B.J. Saradava, Waste heat recovery using Stirling engine. Int. J. Adv. Eng. Technol. **3**, 305–310 (2012)
117. S.S. Mathapati, M. Gupta, S. Dalimkar, A study on automobile air-conditioning based on absorption refrigeration system using exhaust heat of a vehicle. Int. J. Eng. Res. Gen. Sci. **2**(4), 80–86 (2014)
118. S.K. Maurya, S. Awasthi, S.A. Siddiqui, A cooling system for an automobile based on vapour absorption refrigeration cycle using waste heat of an engine. diesel engine **35**, 30–40 (2014)
119. B.I. Ismail, W.H. Ahmed, Thermoelectric power generation using waste-heat energy as an alternative green technology. Recent Patents Electr. Electron. Eng. (Formerly Recent Patents Electr. Eng.) **2**(1), 27–39 (2009)
120. C. Arzbaecher, K. Parmenter, E. Fouche, Industrial waste-heat recovery: Benefits and recent advancements in technology and applications, in *Proceedings of the ACEEE*, 1–2 (2007)
121. The Engineering ToolBox, Fuels Exhaust Temperatures, *The Engineering ToolBox.* (2017)
122. B. Peris, J. Navarro-Esbrí, F. Molés, A. Mota-Babiloni, Experimental study of an ORC (organic Rankine cycle) for low grade waste heat recovery in a ceramic industry. Energy **85**, 534–542 (2015)
123. B.J. Cooper, H.J. Jung, J.E. Thoss, U.S. Patent No. 4,902,487. Washington, DC: U.S. Patent and Trademark Office (1990)
124. N.R. Nwakuba, S.N. Asoegwu, K.N. Nwaigwe, Energy requirements for drying of sliced agricultural products: A review. Agric. Eng. Int. CIGR J. **18**(2), 144–155 (2016)

125. M.A. Billiris, T.J. Siebenmorgen, A. Mauromoustakos, Estimating the theoretical energy required to dry rice. J. Food Eng. **107**(2), 253–261 (2011)
126. G.S.V. Raghavan, T.J. Rennie, P.S. Sunjka, V. Orsat, W. Phaphuangwittayakul, P. Terdtoon, Overview of new techniques for drying biological materials with emphasis on energy aspects. Brazilian J. Chem. Eng. **22**(2), 195–201 (2005)
127. M.H. Masud, M. Nuruzzaman, R. Ahamed, A.A. Ananno, A.A. Tomal, Renewable energy in Bangladesh: current situation and future prospect. International Journal of Sustainable Energy, **39**(2), 132–175 (2020)
128. S. Gunasekaran, T.L. Thompson, Optimal energy management in grain drying. Crit. Rev. Food Sci. Nutr. **25**(1), 1–48 (1986)
129. M. Masud, M.U.H. Joardder, M.T. Islam, M.M. Hasan, M.M. Ahmed, Feasibility of utilizing waste heat in drying of plant-based food materials, in *International conference on mechanical, industrial and materials engineering*, RUET, Rajshahi, 500–503 (2017)
130. C. for E. Cooperation, North American Power Plant Air Emissions (2015)
131. A. Motevali, S. Minaei, M.H. Khoshtaghaza, H. Amirnejat, Comparison of energy consumption and specific energy requirements of different methods for drying mushroom slices. Energy **36**(11), 6433–6441 (2011)
132. M.A. Karim, M.N.A. Hawlader, Performance evaluation of a v-groove solar air collector for drying applications. Appl. Therm. Eng. **26**(1), 121–130 (2006)
133. M.A. Karim, M. Hawlader, Development of solar air collectors for drying applications. Energy Convers. Manag. **45**(3), 329–344 (2004)
134. I.N. Simate, Optimization of mixed-mode and indirect-mode natural convection solar dryers. Renew. Energy **28**(3), 435–453 (2003)
135. F.K. Forson, M.A.A. Nazha, H. Rajakaruna, Modelling and experimental studies on a mixed-mode natural convection solar crop-dryer. Sol. Energy **81**(3), 346–357 (2007)
136. N. Duc Pham et al., Quality of plant-based food materials and its prediction during intermittent drying. Crit. Rev. Food Sci. Nutr. **59**(8), 1197–1211 (2019)
137. N.D. Pham, W. Martens, M.A. Karim, M.U.H. Joardder, Nutritional quality of heat-sensitive food materials in intermittent microwave convective drying. Food Nutr. Res. **62** (2018)
138. M.U.H. Joardder, M. Mourshed, M.H. Masud, Bound water removal techniques, in *State of Bound Water: Measurement and Significance in Food Processing*, Springer, 93–118 (2019)
139. M.I.H. Khan, S.A. Nagy, M.A. Karim, Transport of cellular water during drying: An understanding of cell rupturing mechanism in apple tissue. Food Res. Int. **105**, 772–781 (2018)
140. M.I.H. Khan, T. Farrell, S.A. Nagy, M.A. Karim, Fundamental understanding of cellular water transport process in bio-food material during drying. Sci. Rep. **8**(1), 15191 (2018)
141. M.I.H. Khan, M.U.H. Joardder, C. Kumar, M.A. Karim, Multiphase porous media modelling: A novel approach to predicting food processing performance. Crit. Rev. Food Sci. Nutr. **58**(4), 528–546 (2018)
142. M.U.H. Joardder, C. Kumar, R.J. Brown, M.A. Karim, A micro-level investigation of the solid displacement method for porosity determination of dried food. J. Food Eng. **166**, 156–164 (2015)
143. M.U.H. Joardder, A. Karim, C. Kumar, Effect of temperature distribution on predicting quality of microwave dehydrated food. J. Mech. Eng. Sci. **5**, 562–568 (2013)
144. M.U.H. Joardder, A. Karim, C. Kumar, R.J. Brown, Determination of effective moisture diffusivity of banana using thermogravimetric analysis. Procedia Eng. **90**, 538–543 (2014)
145. C.A. Perussello, C. Kumar, F. de Castilhos, M.A. Karim, Heat and mass transfer modeling of the osmo-convective drying of yacon roots (Smallanthus sonchifolius). Appl. Therm. Eng. **63**(1), 23–32 (2014)
146. M.U.H. Joardder, M.H. Masud, Harmful side effects of food processing, in *Food Preservation in Developing Countries: Challenges and Solutions*, Springer, 153–173 (2019)
147. E. Maltini, D. Torreggiani, G. Bertolo, M. Stecchini, Recent developments in the production of shelf-stable fruit by osmosis, in *Proceedings of 6th International Congress Food Science and Technology*, 177–180 (1983)

148. M. Dalla Rosa, F. Giroux, Osmotic treatments (OT) and problems related to the solution management. J. Food Eng. **49**(2–3), 223–236 (2001)
149. Z. Welsh, C. Kumar, A. Karim, Preliminary investigation of the flow distribution in an innovative intermittent convective microwave dryer (IMCD). Energy Procedia **110**, 465–470 (2017)
150. Z. Welsh, M.J. Simpson, M.I.H. Khan, M.A. Karim, Multiscale Modeling for food drying: State of the art. Compr. Rev. Food Sci. Food Saf. **17**(5), 1293–1308 (2018)
151. C. Kumar, M.A. Karim, M.U.H. Joardder, Intermittent drying of food products: A critical review. J. Food Eng. **121**, 48–57 (2014)
152. C. Kumar, M.U.H. Joardder, T.W. Farrell, M.A. Karim, Investigation of intermittent microwave convective drying (IMCD) of food materials by a coupled 3D electromagnetics and multiphase model. Dry. Technol. **36**(6), 736–750 (2018)
153. C. Kumar, M.U.H. Joardder, T.W. Farrell, G.J. Millar, M.A. Karim, Mathematical model for intermittent microwave convective drying of food materials. Dry. Technol. **34**(8), 962–973 (2016)
154. C. Kumar, M.U.H. Joardder, A. Karim, G.J. Millar, Z. Amin, Temperature redistribution modelling during intermittent microwave convective heating. Procedia Eng. **90**, 544–549 (2014)
155. C. Kumar, M.U.H. Joardder, T.W. Farrell, M.A. Karim, Multiphase porous media model for intermittent microwave convective drying (IMCD) of food. Int. J. Therm. Sci. **104**, 304–314 (2016)
156. C. Kumar, M.U.H. Joardder, T.W. Farrell, G.J. Millar, A. Karim, A porous media transport model for apple drying. Biosyst. Eng. **176**, 12–25 (2018)
157. C. Kumar, M.A. Karim, Microwave-convective drying of food materials: A critical review. Crit. Rev. Food Sci. Nutr. **59**(3), 379–394 (2019)
158. C. Kumar, G.J. Millar, M.A. Karim, Effective diffusivity and evaporative cooling in convective drying of food material. Dry. Technol. **33**(2), 227–237 (2015)
159. H. Feng, Y. Yin, J. Tang, Microwave drying of food and agricultural materials: Basics and heat and mass transfer modeling. Food Eng. Rev. **4**(2), 89–106 (2012)
160. M.U.H. Joardder, C. Kumar, M.A. Karim, Multiphase transfer model for intermittent microwave-convective drying of food: Considering shrinkage and pore evolution. Int. J. Multiph. Flow **95**, 101–119 (2017)
161. I.W. Turner, P.C. Jolly, Combined microwave and convective drying of a porous material. Dry. Technol. **9**(5), 1209–1269 (1991)
162. A. Datta, V. Rakesh, *An Introduction to Modeling of Transport Processes: Applications to Biomedical Systems,* Cambridge University Press (2010)
163. M. Zhang, J. Tang, A.S. Mujumdar, S. Wang, Trends in microwave-related drying of fruits and vegetables. Trends Food Sci. Technol. **17**(10), 524–534 (2006)
164. I.W. Turner, J.R. Puiggali, W. Jomaa, A numerical investigation of combined microwave and convective drying of a hygroscopic porous material: A study based on pine wood. Chem. Eng. Res. Des. **76**(2), 193–209 (1998)
165. A.S. Mujumdar, *Handbook of Industrial Drying,* CRC press (2006)
166. M. Zhang, H. Jiang, R.-X. Lim, Recent developments in microwave-assisted drying of vegetables, fruits, and aquatic products—Drying kinetics and quality considerations. Dry. Technol. **28**(11), 1307–1316 (2010)
167. S. Gunasekaran, Pulsed microwave-vacuum drying of food materials. Dry. Technol. **17**(3), 395–412 (1999)
168. M.I.H. Khan, M.A. Karim, Cellular water distribution, transport, and its investigation methods for plant-based food material. Food Res. Int. **99**, 1–14 (2017)
169. M.I.H. Khan, C. Kumar, M.U.H. Joardder, M.A. Karim, Determination of appropriate effective diffusivity for different food materials. Dry. Technol. **35**(3), 335–346 (2017)
170. M.I.H. Khan, R.M. Wellard, S.A. Nagy, M.U.H. Joardder, M.A. Karim, Experimental investigation of bound and free water transport process during drying of hygroscopic food material. Int. J. Therm. Sci. **117**, 266–273 (2017)
171. M.I.H. Khan, R.M. Wellard, S.A. Nagy, M.U.H. Joardder, M.A. Karim, Investigation of bound and free water in plant-based food material using NMR T2 relaxometry. Innov. Food Sci. Emerg. Technol. **38**, 252–261 (2016)

172. L. Huang, M. Zhang, L. Wang, A.S. Mujumdar, D. Sun, Influence of combination drying methods on composition, texture, aroma and microstructure of apple slices. LWT-Food Sci. Technol. **47**(1), 183–188 (2012)
173. Y. Wang, M. Zhang, A.S. Mujumdar, K.J. Mothibe, S.M.R. Azam, Effect of blanching on microwave freeze drying of stem lettuce cubes in a circular conduit drying chamber. J. Food Eng. **113**(2), 177–185 (2012)
174. A. Andrés, C. Bilbao, P. Fito, Drying kinetics of apple cylinders under combined hot air–microwave dehydration. J. Food Eng. **63**(1), 71–78 (2004)
175. L. Cinquanta, D. Albanese, A. Fratianni, G. La Fianza, M. Di Matteo, Antioxidant activity and sensory attributes of tomatoes dehydrated by combination of microwave and convective heating. Agro Food Ind Hi Tech **24**(6), 35–38 (2013)
176. D.G. Prabhanjan, H.S. Ramaswamy, G.S.V. Raghavan, Microwave-assisted convective air drying of thin layer carrots. J. Food Eng. **25**(2), 283–293 (1995)
177. D. Argyropoulos, A. Heindl, J. Müller, Assessment of convection, hot-air combined with microwave-vacuum and freeze-drying methods for mushrooms with regard to product quality. Int. J. Food Sci. Technol. **46**(2), 333–342 (2011)
178. W. Jindarat, P. Rattanadecho, S. Vongpradubchai, Y. Pianroj, Analysis of energy consumption in drying process of non-hygroscopic porous packed bed using a combined multi-feed microwave-convective air and continuous belt system (CMCB). Dry. Technol. **29**(8), 926–938 (2011)
179. Y. Soysal, M. Arslan, M. Keskin, Intermittent microwave-convective air drying of oregano. Food Sci. Technol. Int. **15**(4), 397–406 (2009)
180. H. Jiang, M. Zhang, A.S. Mujumdar, R. Lim, Comparison of drying characteristic and uniformity of banana cubes dried by pulse-spouted microwave vacuum drying, freeze drying and microwave freeze drying. J. Sci. Food Agric. **94**(9), 1827–1834 (2014)
181. K.J. Mothibe, C.-Y. Wang, A.S. Mujumdar, M. Zhang, Microwave-assisted pulse-spouted vacuum drying of apple cubes. Dry. Technol. **32**(15), 1762–1768 (2014)
182. Y. Wang, M. Zhang, A.S. Mujumdar, K.J. Mothibe, Microwave-assisted pulse-spouted bed freeze-drying of stem lettuce slices—Effect on product quality. Food Bioprocess Technol. **6**(12), 3530–3543 (2013)
183. Y. Soysal, Z. Ayhan, O. Eştürk, M.F. Arıkan, Intermittent microwave–convective drying of red pepper: Drying kinetics, physical (colour and texture) and sensory quality. Biosyst. Eng. **103**(4), 455–463 (2009)
184. G.E. Botha, J.C. Oliveira, L. Ahrné, Microwave assisted air drying of osmotically treated pineapple with variable power programmes. J. Food Eng. **108**(2), 304–311 (2012)
185. O. Esturk, Intermittent and continuous microwave-convective air-drying characteristics of sage (Salvia officinalis) leaves. Food Bioprocess Technol. **5**(5), 1664–1673 (2012)
186. O. Esturk, M. Arslan, Y. Soysal, I. Uremis, Z. Ayhan, Drying of sage (Salvia officinalis L.) inflorescences by intermittent and continuous microwave-convective air combination. Res. Crop. **12**(2), 607–615 (2011)
187. V. Orsat, W. Yang, V. Changrue, G.S.V. Raghavan, Microwave-assisted drying of biomaterials. Food Bioprod. Process. **85**(3), 255–263 (2007)
188. L.M. Ahrné, N.R. Pereira, N. Staack, P. Floberg, Microwave convective drying of plant foods at constant and variable microwave power. Dry. Technol. **25**(7–8), 1149–1153 (2007)
189. Y. Nishiyama, W. Cao, B. Li, Grain intermittent drying characteristics analyzed by a simplified model. J. Food Eng. **76**(3), 272–279 (2006)
190. S. Gunasekaran, H.-W. Yang, Effect of experimental parameters on temperature distribution during continuous and pulsed microwave heating. J. Food Eng. **78**(4), 1452–1456 (2007)
191. S. Gunasekaran, H.-W. Yang, Optimization of pulsed microwave heating. J. Food Eng. **78**(4), 1457–1462 (2007)
192. Y. Soysal, Intermittent and continuous microwave-convective air drying of potato (lady rosetta): drying kinetics, energy consumption and product quality. Tarım Makinaları Bilimi Dergisi, **5**(2), 139–148 (2009)

193. D. Zhao, K. An, S. Ding, L. Liu, Z. Xu, Z. Wang, Two-stage intermittent microwave coupled with hot-air drying of carrot slices: Drying kinetics and physical quality. Food Bioprocess Technol. **7**(8), 2308–2318 (2014)
194. N. Aghilinategh, S. Rafiee, S. Hosseinpour, M. Omid, S.S. Mohtasebi, Optimization of intermittent microwave–convective drying using response surface methodology. Food Sci. Nutr. **3**(4), 331–341 (2015)
195. J. R. de J. Junqueira, J.L.G. Corrêa, D.B. Ernesto, Microwave, convective, and intermittent microwave–convective drying of pulsed vacuum osmodehydrated pumpkin slices. J. Food Process. Preserv. **41**(6) (2017)
196. O.M. Kesbi, M. Sadeghi, S.A. Mireei, Quality assessment and modeling of microwave-convective drying of lemon slices. Eng. Agric. Environ. Food **9**(3), 216–223 (2016)
197. N.D. Pham, C. Kumar, M. Joardder, H. Khan, W. Martens, M.A. Karim, Effect of Different Power Ratio Mode of Intermittent Microwave Convective Drying on Quality Attributes of Kiwi Fruit Slices, 7–10, August (2016)
198. M.U.H. Joardder, M.H. Masud, S. Nasif, J.A. Plabon, S.H. Chaklader, Development and performance test of an innovative solar derived intermittent microwave convective food dryer, in *AIP Conference Proceedings*, **2121**(1), 40010–40013 (2019)
199. M.K. Krokida, V.T. Karathanos, Z.B. Maroulis, D. Marinos-Kouris, Drying kinetics of some vegetables. J. Food Eng. **59**(4), 391–403 (2003)
200. A. Kaya, O. Aydin, C. Demirtas, M. Akgün, An experimental study on the drying kinetics of quince. Desalination **212**(1–3), 328–343 (2007)
201. S.J. Babalis, V.G. Belessiotis, Influence of the drying conditions on the drying constants and moisture diffusivity during the thin-layer drying of figs. J. Food Eng. **65**(3), 449–458 (2004)
202. K. Sacilik, A.K. Elicin, The thin layer drying characteristics of organic apple slices. J. Food Eng. **73**(3), 281–289 (2006)
203. M.J. Barroca, R. Guiné, Study of drying kinetics of quince, in *International Conference of Agricultural Engineering CIGR-AgEng* (2012)
204. D.A. Tzempelikos, A.P. Vouros, A.V. Bardakas, A.E. Filios, D.P. Margaris, Case studies on the effect of the air drying conditions on the convective drying of quinces. Case Stud. Therm. Eng. **3**, 79–85 (2014)
205. A.A. Ananno, M.H. Masud, P. Dabnichki, A. Ahmed, Design and numerical analysis of a hybrid geothermal PCM flat plate solar collector dryer for developing countries. Solar Energy **196**, 270–286 (2020)
206. M.H. Masud, A.A. Ananno, N. Ahmed, P. Dabnichki, K.N. Salehin, Experimental investigation of a novel waste heat based food drying system. Journal of Food Engineering **281**,110002 (2020)
207. M.U. Joardder, M.H. Masud, S, Nasif, J.A. Plabon, S.H. Chaklader, Development and performance test of an innovative solar derived intermittent microwave convective food dryer. In AIP Conference Proceedings. **2121**(1)1, 040010, July (2019)

Chapter 6
Challenges in Implementing Proposed Sustainable Food Drying Techniques

Food industry and consumer culture have evolved considerably over the past few decades. Earlier, a simple heat and mass transfer mechanisms were used for extending the shelf life of dried food products. However, as the knowledge about the dried food properties is becoming increasingly available, research and development on drying technologies have expanded beyond the limited mechanical and chemical engineering approach of heat and mass transfer. As a result, improved and efficient drying technologies are not a future dream anymore; it is within reach of people in developing countries. The recent trend is to produce dried food products, which can retain various qualities of fresh food, including texture, nutrition, sensory properties, flavors, and color [1, 2] (see Chap. 2 for details). Modern dryers are expected to have the capacity to produce high quality dried food as well as accomplish the drying process in an energy-efficient and inexpensive way. Chapter 5 discussed and recommended potential drying techniques that can meet the drying industry benchmark and produce improved dry food. In this chapter, the challenges associated with adopting the recommended drying techniques are discussed from the perspective of developing countries. In most cases, the fundamental challenge behind high quality dried food production is to determine the optimum drying condition (operating conditions) for a particular drying technology [3, 4]. A general view of the effects of process severity on the food quality is presented in Fig. 6.1 [5].

When designing energy-efficient, cost-effective, and sustainable drying techniques for developing countries, engineers and researchers should concentrate their focus on challenges associated with drying operation, post drying storage, packaging system, and retention of food quality. Careful evaluation should be done on the financial, technical, and policy-related issues that might occur if a developing country chooses to invest in an improved drying technique. Therefore, this chapter aims to identify the bounds of applicability for the proposed improved drying techniques discussed in Chap. 5. It is expected that future research would one day be able to solve these challenges.

© Springer Nature Switzerland AG 2020
M. Hasan Masud et al., *Sustainable Food Drying Techniques
in Developing Countries: Prospects and Challenges*,
https://doi.org/10.1007/978-3-030-42476-3_6

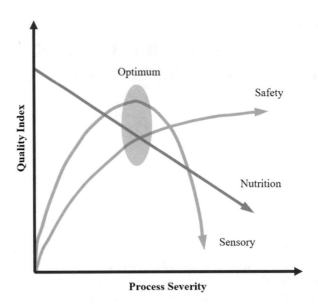

Fig. 6.1 Effects of process severity of food quality

Additionally, the opposing nature of cost feasibility and energy efficiency imposes a complex challenge. The market available drying techniques are two types: the first types have low efficiency, is simple in design, require low-power but is very cost-effective. The other types of the dryer are highly energy-efficient, have long-life but are very expensive dryers. Thus, finding a trade-off between energy efficiency and cost-effectiveness is a challenge. This book, therefore, proposed renewable energy-based improved drying technologies to prioritize energy efficiency in favor of environmental sustainability. This is in part due to the exorbitant cost associated with the climate change crisis.

6.1 Overall Challenges of Drying

Dehydrated foods can appeal to a wide range of consumers irrespective of age and ethnicity. Long shelf life, lower processing cost, and zero preservatives are some of the major benefits enjoyed by dry food enthusiasts [6]. Drying can be done in numerous ways depending on the desired level of food quality, expected drying time, available sources of energy, and type of food material [7, 8] (See Chaps. 4 and 5 for more information). Despite the technical soundness and various benefits, certain challenges are associated with drying systems and processes. Before the specific challenges are being discussed, the general bounds of applicability, which is commonly associated with most dryers, are presented here.

6.1.1 Preparation of Food Samples

Before drying can be accomplished, it is essential to slice the drying fruits and vegetables in an appropriate shape and size. If the samples are too thin, it will dry considerably faster, but there is a risk that its nutritional values, texture, and taste will be lost [3, 4]. In contrast, if food sample slices are too thick, it will take a considerably long time to dehydrate. Longer drying time will increase the cost of the finished product, and the inside of food may not dry properly. Additionally, the dryness level is significantly affected by the volume of food that is added in a single batch. Determining the optimal amount of food, which is to be dried in a single session, can be challenging for general consumers.

Typically, users resort to the process of trial and error to determine the right shape, size of the sample, and amount of the vegetable and fruits to dry in a batch. Furthermore, a lack of knowledge of the operating procedures of an improved dryer might also cause improper drying.

6.1.2 Environmental Cleanliness

While selecting the type of drying technique in order to process a particular food, cleanliness and hygiene should be given top priority. Keeping the dryer clean and sanitized is imperative to producing a high-quality finished product. Over time, the repeated use of a dryer contaminates its drying chamber. Small food particles that get stuck inside the drying trays or small perforated holes can be a source for microbial growth. If the dying chambers are not cleaned after certain intervals, the microbial growth can increase exponentially. In that case bacteria, dust, and debris can deteriorate the dried food quality significantly. Additionally, the operator should use clean utensils for handling the food materials or wash their hands carefully before and after food handling.

Dried food products that are prepared in an unclean contaminated environment are susceptible to bacterial infection. Consumption of such contaminated food might poison the user. Therefore, a significant concern should be given to environmental cleanliness.

6.1.3 Need for Specialized Equipment

In many dryers' special equipment and chemicals are needed to be restocked. Sensitive equipment might malfunction after a number of drying operation and chemicals such as phase change materials, and hypotonic solutions should be resupplied once their stock is over. Therefore, it is essential to calculate the depreciation cost of specialized drying equipment to avoid operational failure.

Additionally, if drying is to be done regularly, it is essential to store the processed food in a safe and contamination-free environment. In developing countries with high humidity, equipment such as dehumidifier and vacuum sealer should be used to make sure the dried food samples do not re-absorb moisture from the atmosphere.

6.1.4 Food Texture

For any dryer, it is difficult to maintain consistency and precise dryness fraction in successive drying operation. Often it is challenging to gain the expected dryness level as slight changes in the drying parameters might produce different quality finished products. Food materials that are dried with high temperatures might crumble if touched. Moreover, uneven heat distribution inside the dryer might form a curst on the side of the food as well as an inconsistent dried surface.

6.1.5 Drying Time

Despite the advancement in the dryer, the drying process is considerably time-consuming. Few drying techniques which offer faster drying are very expensive and difficult to operate. In most dryers, it might take several hours before the food material is sufficiently moisture-free from inside and ready for storage.

6.1.6 Energy Required for Drying

Operating a dryer regularly for several hours a day will undoubtedly consume an enormous amount of energy. If the dryer operates on fossil fuel or fossil-based electricity, it will undoubtedly increase the cost of drying. A considerable percentage of energy in the food processing industry is consumed by the drying and heating process; therefore, the energy factor should be carefully considered [9, 10]. That is to say, for short term benefit of low-cost fossil-based drying techniques should not be prioritized over renewable energy-based high initial investment drying systems. While the fossil-based dryer can operate with inexpensive initial installment costs, due to high operating costs over time, it will prove to be more expensive than any form of renewable energy-based drying techniques. Optimizing the energy efficiency of drying techniques have always been a challenge [11].

6.1.7 Nutrition Content

The quality of dried food has improved significantly over the years due to the improvement of drying techniques. However, no drying techniques exist in the market, which can retain the original nutritional content of fresh food once it is dried. If the drying is done improperly, a significant percentage of the vitamins and minerals of the food sample may be irreversibly damaged.

Furthermore, some dried food becomes rich in salt and sugar once the drying operation is complete. If the salt and sugar concentration is not maintained correctly, once consumed, it might raise the blood pressure of the individual. It has been observed that the salt concentration of dried snacks is above the recommended salt intake level of adults. Drying technologies should be designed to maintain an adequate level of nutrition inside the dried food samples, which is always proved as one of the challenging tasks.

6.1.8 Taste of Food

Often the dried food is too hard or requires moistening up before consumption. In those cases, the consumer might dislike dried food. Improper drying may also damage the soft and tenderness aspects of fruits and vegetables, which is unappealing to many consumers. Therefore, it is vital to ensure that the consumer finds dried food tasty and attractive. Drying techniques should be selected considering the taste factor.

6.2 Specific Challenges in Implementing Proposed Drying Techniques

Chapter 5 has recommended improved drying technologies for developing countries based on specific selection criteria such as drying quality, energy, time, and cost (see Chap. 2 for details). Drying involves heating of drying medium, and in the drying process, a huge amount of energy is lost as it is impossible to utilize 100% of the heat energy. Therefore, all dryers are susceptible to limitations and disadvantages. Specific challenges related to the implementation, maintenance, and operation of the proposed drying techniques are presented critically here.

6.2.1 Challenges in Implementing Improved Solar Drying Systems

Despite the renewable and non-polluting nature of solar energy, it is difficult to use solar radiation for drying. While there is an abundance of solar energy, its periodic nature contradicts the idea of continuous operation. Therefore, the inherent problem of all solar-based technologies is that they are ineffective during off-sun shine hours. Although this issue can be addressed with the use of a thermal storage system, but they have their own list of limitations. Alongside this thermal storage, the difficulty of solar energy can be mediated using auxiliary energy sources. However, that will increase the cost of the entire drying system.

Moreover, since the incident radiation intensity is a function of time, the optimal intensity is available only for 1–3 hours each day. During the morning and afternoon hours, the solar intensity is considerably low. Another glaring problem of solar-based technologies is the low energy density of solar radiation. Because of low energy density, solar dryers require a large collector area, which may be expensive. Considering the fact that fossil-based technologies can supply enormous heat flux using a compact setup, it is impractical for solar dryers to demand large energy-collecting surfaces.

The application of solar drying technologies is also limited by the geographic position of a particular country. Generally, developing countries are situated in the tropical region, which is suitable for solar drying due to high global horizontal solar irradiance. Developing nations situated beyond the tropic of Cancer and the tropic of Capricorn can use solar radiation on a limited scale. Moreover, the performance of solar dryers is also susceptible to seasonal changes, meaning during winter or rainy season, the solar dryers are extremely ineffective. Hence, it is essential to select seasonal crops for solar dryers to govern investment costs. Additionally, only a limited number of food products can be dried using cheaper flat-plate solar collectors, which can heat the air up to 60 °C. High-temperature drying can only be achieved through concentrated solar collectors. However, these types of dryers are considerably expensive. Given the small energy flux of most solar dryers, they can only be used for drying food, which requires low energy.

The congenital problems of solar dryers can be summarized as (i) they require expensive heat storage systems or auxiliary energy source or both (ii) they lack a temperature control system, and (iii) in order to operate efficiently, they need large-surface collectors. The solutions to these problems necessitate high investment, which may be a challenge for the low-income developing countries of the world.

6.2.2 Challenges in Implementing Geothermal Drying Systems

Challenges associated with the adoption of geothermal drying in developing countries can be classified as: (i) financial challenges; (ii) policy and regulation challenges; and (iii) technical challenges. The cause and effects of these challenges are presented critically here.

6.2.2.1 Policy and Regulation

- Although geothermal energy is an accepted form of renewable energy, very few governments of the developing countries have clear policies and legislations for its utilization. Due to the lack of interest of the government, domestic investors and foreign private sectors are reluctant to invest in geothermal energy-based drying.
- Geothermal energy requires a high initial investment and costly infrastructure development. Such financial capacity is unavailable to most developing countries. Moreover, due to inadequate legislative framework, it is challenging to receive funds for geothermal surveys and exploration, which can potentially lower the cost of its utilization. Without the right policy and early phase financial support from the government for research and appraisal, it would very difficult to adopt geothermal dryers in developing countries.
- Most developing countries lack political stability, institutional framework, and consolidation among stakeholders. These are pre-requisite for establishing geothermal energy and drying industries.

6.2.2.2 Technical Barrier

- The adoption of geothermal technologies (e.g., geothermal dryer) requires high-level technical expertise. However, most developing countries suffer from a shortage of qualified geothermal engineers, economic managers, trained labors, and policy analysts to successfully implement a geothermal project. Most geothermal dryers work in association with a geothermal power plant. The rarity of geothermal plants in developing countries is also a significant challenge.
- Developing countries with less advanced communication networks, organizational culture, and transportation systems will find it difficult to adopt geothermal drying technology.

6.2.2.3 Financial Barrier

- One of the biggest challenges of implementing geothermal drying technology is its exorbitant high upfront cost. This situation is exacerbated by the lack of financial incentives (e.g., green tariffs, feed-in tariff) for adopting renewable energy technologies. These renewable energy sources are inherently more expensive than fossil fuels. Therefore, without governmental support and tax break policies, it will be very difficult for private companies to finance renewable energy technology such as a geothermal dryer. It is difficult for low- and middle-income developing countries to distribute their limited budget among different development sectors (e.g. health, education, national security). As a result, the growth of geothermal drying technology might face difficulties.
- Since most low and middle-income developing countries do not yet have a net metering policy, it is a challenge to find foreign investors for financing expensive renewable energy projects. Moreover, without any specific tax break, consumers will not be interested in paying higher for dried food produced using renewable energy. The unfavorable terms and conditions of mutual agreements also drive away potential investors.

6.2.3 Challenges in Implementing Hybrid Geothermal Drying Systems

Scientific research and engineering innovation have given the consumer several ingenious techniques to better utilize the immense solar potential. However, the integration of said technologies with the existing energy framework can be challenging and often requires extensive market research. There are some challenges that are to be faced with implementing this hybrid system. These challenges are described below:

6.2.3.1 Specific Challenges

In the case of Solar Air Heater,

- Air has poor heat transfer properties. Collectors need to be faced towards the sun. Thus, the requirement of the proper sun-tracking facility makes the system challenging.
- The complex geometry of the collector (e.g. V-groove, finned) is required to increase efficiency, but it increases the cost.

In the case of PCM,

- Proper heat distribution inside the PCM compartment is difficult in a practical setup, which reduces the efficiency

- Evaporation of paraffin and its leakage to the environment may lead to economic loss
- The compartment should be stable and compatible to store PCM
- Encapsulation increases the efficiency of PCM, but increased complexity and cost limits its use.

In the case of geothermal energy,

- Geothermal energy collection requires a high initial investment
- Depending on the depth, costly setup is needed, and high power is necessary to make the fluid flow in the geothermal heat exchanger
- Rigorous geological survey is required to find the potential areas for geothermal energy
- Sulphur dioxide, silica, and toxic heavy metal like arsenic, mercury, boron, etc. emission come with geothermal energy.

6.2.3.2   Geographical Challenge

Both the solar irradiance and geothermal energy depends on the geography of the place. It might be challenging for developing countries to locate an ideal place where high solar irradiation is accompanied by high geothermal energy without a comprehensive survey. Besides, the solar collector is required to face the sunlight for most of the time, which makes the system application difficult. Hybrid geothermal dryers will be economically practical and useful, provided that only if the average annual sunshine time (approximately greater than 2600 hr) is high and the yearly total quantity of radiation is enough (more than 6×10^6 KJ/m^2) [6].

6.2.3.3   Technical Challenge

Despite the theoretical potential of supplying higher than ambient air temperature for 24 hours a day, the hybrid geothermal PA-FPSC requires electric power to operate its pumps and blowers. In an ideal condition, the pumps and associated electrical systems can run on PV cell technology. However, the average sun hour in developing countries is about 7 hours; hence, after dark, it would be difficult to support the system without appropriate electric power backup. The limitation can be avoided with the application of an electronic battery system, although that would increase the cost of the system. Maintaining the geothermal earth to the air heat exchanger is also difficult.

6.2.4   Challenges in Utilising Heat Pump Drying Systems

Heat pump dryers' advantages clearly indicate that they represent a smart investment in the long run. Limitations of the heat pump dryer are described below:

- Chlorofluorocarbons (CFCs) are used in refrigerant cycles, some of which are not environment friendly. It requires regular maintenance of components (e.g., compressor and refrigerant filters) and charging of the refrigerant, which may incur a higher capital cost [12].
- This process has a lower heat supply compared to oil and gas boilers, so larger radiators would be needed for high-temperature drying. These systems can take significant time to heat up. Additionally, the scheme of collecting the required energy from solar or wind power, may not be a sustainable solution given that they have their own limitations. Without proper insulation, system efficiency will drop significantly.
- These systems are noisy, which is made worse during the winter seasons. Since heat pump dryers have a low Coefficient of Performance (COP), they have low efficiency during winter. As a result, to achieve effective drying, heat pump dryers need to run continuously during winter. This increases the operation cost on top of its high upfront installment cost. The installation cost of advanced heat pump dryers can be as high as 24,500 USD, which is very difficult to afford for low and middle-income developing countries [13].
- Heat pump systems are challenging to install and maintain. Troubleshooting of the heat pump system is not possible without trained professionals. Additionally, the heat transfer fluid used in this system has questionable sustainability. If exposed, they can cause a detrimental threat to the environment. Therefore, companies should use biodegradable fluids.
- Since heat pump dryers require electricity, unless they are operating on renewable energy, these dryers will have a high carbon footprint. Additionally, over-reliance on fossil fuel energy will rack up the operating cost significantly. However, the proposed heat pump dryers with solar integration can effectively produce a zero-energy heat pump dryer. Due to the high implementation cost, it would only be beneficial to middle- and high-income developing countries. In some countries, because of the environmental production laws and regulations, planning permissions might be required for installing heat pumps.

6.2.5   Challenges in Applying Waste Heat Convective Drying (WHCD) Systems

The proposed innovative WHCD system can utilize the waste heat recovered from different sources, including combustion engines, generators, boilers, furnaces, etc. and reuses that heat to dry food for preservation purposes. The proposed process does not require external energy to drive the drying process; therefore, it can reduce entropy and CO_2 emissions. However, the implementation of waste heat convective dryer is susceptible to numerous challenges that are discussed below:

- Fouling Factor will play a significant role in the performance of such dryers. When the WHCD system continues to run for an extended period, carbon particles from the exhaust flue gas may slowly accumulate on the inside surface of the heat exchanger. The fouling factor will increase as the layer of carbon particle thickens. Subsequently, the heat transfer coefficient and overall thermal efficiency of the system will decrease. Life cycle analysis could able to help to determine the maintenance procedure for reducing the fouling factor.
- Air filtration system at the opening of the heat exchanger and the food dryer can improve the quality of dried food. Without proper filtration measures, exhaust flue gas might mix with the drying air and contaminate the food material. The toxicity of the exhaust flue gas particles has a detrimental effect on human health. Therefore, selecting the appropriate filter in order to overcome this challenging issue should be considered while implementing this type of proposed dryer
- Design and integration of heat exchangers for different waste heat sources is a challenging task. Selecting an appropriate material for a specific type of engine or generator is also a complicated task. Because all the sources provide waste heat with a varied range of temperatures. Therefore, a wide range of modelling parameters should be considered. Heat resistive materials are required to design heat exchanger for high-temperature exhaust gas. These materials are expensive and difficult to obtain.
- The initial cost could be higher as different parts, including a PV cell, are costly. Therefore, the payback period could be high, which may be difficult to afford for low-income countries

6.2.6 Challenges in Applying Desiccant Drying Systems

Desiccant system was built and tested by Lof in 1955 at Solar Energy Research Institute, University of Wisconsin [14]. It used liquid triethylene glycol in order to heat the air directly inside a solar collector as well as to re-generate the desiccant. The disadvantage of this system was the leakage of desiccant into the cooling space. When desiccants with corrosive nature were used, it proved to be hazardous to the human operator and material near it. Also, such difficulties render it totally unusable in domestic appliances.

Additional limitations include:

- High initial setup cost of the system is a challenge for the developing countries. Experienced professionals are required to construct, install, and service such systems. Liquid desiccants could be corrosive and could damage the system components.

- These are cost-effective only when there is a source of waste heat available to regenerate the desiccant. Regeneration system is complicated to build and maintain. Limited application in high humidity climate areas. System effectiveness, to a large extent, depends on the desiccant properties. Desiccant materials are costly, and to utilize the immense solar potential better crystallization of aqueous salts is an expensive system with high level of complexity.
- When using liquid desiccants such as lithium bromide, lithium chloride, and other salts that are corrosive, the operators should be extremely cautious not to damage the desiccant dryer. If the desiccant and exhaust air are exposed to the environment without proper filtration, it would be detrimental to human health.
- High power pumps are required to handle and operate large quantity of liquid desiccant, which increases the operation cost of desiccant drying beyond the reach of low-income developing countries.

6.2.7   Challenges in Applying Improved Biomass Drying Systems

Implementing an economically feasible and environmentally sustainable biomass drying technology can be a challenging task [15]. In this sub-section, a summary of the challenges associated with biomass drying is represented.

6.2.7.1   Thermochemical Conversion

Biomass feeds are thermochemically converted (e.g., combustion and gasification) into usable thermal energy for the dryer, which has some shortcomings. Wood chips are primarily used as biomass fuel for combustion-based dryers, which raises major environmental concerns due to carbon pollution [16, 17]. Additionally, excessive use of wood chips may promote deforestation. Unplanned irradiation of natural resources will disrupt the ecosystem and damage biological diversity.

6.2.7.2   Physical Conversion Techniques

Improved cooking stoves and biomass briquetting widely used for biomass combustion inside the dryer. When biomass is burnt in these dryers, an enormous amount of NO_x and particulate matter are released into the environment. These toxic particles have adverse effects on the ecosystem, including acid rain, photochemical smog, and disruption of crop production [18]. Furthermore, the physical conversion of biomass inside the dryer may release by-products such as sulphur oxides (SO_2, SO_3), which have detrimental effect on environmental sustainability [19, 20].

6.2.7.3   Operational Challenges

- The generation of feed material for biomass dryers requires valuable resources like water, land, and energy. Without proper management of these resources and governmental financial support, it is challenging to grow the feed material industry. With the growing demand for drying technologies, if the feed material supply does not increase, the biomass dryers would not be able to function economically.
- Transporting wet biomass is a challenge because of its high moisture content. The presence of moisture not only makes it difficult to handle but also increases its weight. The growth of microbes and odor also makes it unfavorable for storing near food processing industry.
- Inefficient design of biomass combustor produces low thermal energy, which may improperly dry the food products. Conversely, efficient combustors are difficult to maintain and often expensive. Moreover, the lack of appropriate pre-treatment process accelerates biodegradation and increases the loss of heating value. As a result, the production and equipment maintenance cost of biomass dryer increases.
- Due to lack of communication, it has proven to be very difficult to keep long term sustainable contracts with biomass feedstock suppliers at an acceptable price. The minimum profitability of the biomass feed industry makes it difficult to attract private companies to support the upstream firms for producing biomass feed for the dryer.

6.2.7.4   Economic Challenges

- Since the biomass resources are scattered in small regions isolated from each other. It is expensive to transport biomass feed from one place to another. Most dryers are installed along with food processing industries; therefore, high transportation costs involved in ensuring a constant supply of biomass feed will be challenging.
- Owing to lack of capital, subsidies on feed production, inadequate profitability accompanied by a low number of investors, and high market risk make it very challenging to invest in biomass drying.

6.2.7.5   Policy and Regulation

- In many developing countries, there are no concrete regulations or financial incentives for utilizing biomass energy. Therefore, companies do not have an economic interest in investing in biomass energy-based dryers.
- Due to the lack of a special mechanism for managing and developing the biomass resources industry, it is difficult to find the biomass feeds at low prices.

Since most developing countries do not have a feed-in tariff or green tariff on biomass energy, this industry is severely underdeveloped.

6.2.8   Challenges in Applying Hybrid Solar-Biomass Drying Systems

Challenges associated with hybrid solar-biomass drying is a combination of the challenges discussed in the solar and biomass drying. Since this is a hybrid system combining the potential of solar and biomass energy, some advantages may ameliorate the disadvantages of others. As such, the potential challenges associated with this dryer is discussed as follows [21]:

- Solar energy and biomass energy both require a large amount of land to establish its infrastructure, which is difficult to attain in developing countries with high population density. Additionally, a constant supply of feed materials should be ensured for the continuous operation of this dryer. Feed materials are available in rural areas; hence, these dryers cannot be installed in city or industrial area.
- Transportation of feed material may increase the operational cost of this dryer. Large-scale application of this dryer might contribute to deforestation and subsequently damage the environment. Production of low moisture high energy-dense Agri-pellets is costly.
- In seasons where the solar energy is unavailable, the hybrid dryer will run entirely on biomass energy. As a result, the operating cost will increase significantly. Moreover, long term use of biomass feeds might deplete soil fertility and nutrition, which might also damage biodiversity and disrupt the ecosystem.
- If the governments of certain developing countries do not stop subsidizing the fossil fuel industry, renewable energy-based technologies will find it difficult to penetrate the market. Since the cost of electricity is low, consumers and food industries will resort to using cheap fossil-based electricity for drying purposes.

6.2.9   Challenges in Applying PMSD Systems

Pulsed Microwave Solar Dryer (PMSD) is introduced with an aim to develop an efficient and economically viable drying system, which can operate using renewable energy sources (e.g. solar energy). Although the system has a high degree of compactness and flexibility, there are some specific challenges associated with the implementation of PMSD. An overview of the challenges associated with the PMSD is given here:

- The major challenge in any microwave system is the non-uniformity of the electric field [22, 23]. Without a uniform electric field distribution, the drying temperature and moisture distribution will not be uniform, which will produce low quality dried food.

- Optimizing the pulse ratio for different food materials is very difficult as the different food materials have their unique characteristics [24, 25]. Furthermore, there are numerous ways by which the pulse ratio of the microwave can be selected, and choosing the exact on-time and off-time of microwave is a challenging task. Moreover, there is a wider variety of process conditions and food properties, selection of these conditions and properties while optimizing the microwave is also a tough task. It is often challenging to maintain a satisfactory quality of the food due to improper selection of pulse ratio that leads to crust formation [25, 26]. Moreover, samples may burn in different places due to the non-uniformity of electric field.
- There will have considerable risk for microwave leakage that will subsequently affect the uniformity of the electric field. High-intensity microwave radiation might cause physical damage to the operator; therefore, proper protection should be used.
- If the system runs on solar energy, it would be difficult to operate during the off-sunshine hours without expensive battery backup. Moreover, PMSD is a costly technology that requires a high initial investment, as it consumes high voltage electricity, which is expensive.

6.2.10   Challenges in Applying Osmotic Dehydration Systems

It offers considerable potential for energy saving in comparison with conventional drying. However, most researches on the energy consumption and efficiency of osmotic dehydration were done or analyzed under laboratory conditions [27–30]. These parameters might change when the drying technique is used as part of the large-scale production industry. Therefore, the effectivity of osmotic dehydration in practical scenario might face the following challenges [31]:

- The Osmo-active solutions require constant heating to regenerate its dehydration capacity. Supplying constant heat may require additional investment and operators. The cost of additional heat energy will increase its operating cost of the dryer.
- A complicated control system is required to ensure proper solution mixing and recirculation. It is difficult to operate the control system of the dryer without the help of an expert.
- After the successive operation, the hypertonic substance dissolves and becomes diluted, which may produce low quality dried products after certain operation cycles.
- An expensive evaporator is required for appropriate evaporation of water, which will considerably increase the instalment cost of the dryer

Being constrained by the laws of thermodynamics, it is near impossible to create a drying system with Carnot's efficiency. Therefore, all proposed dryers are subjected to various forms of physical and implementation related challenges. Improved drying techniques are recommended despite their respective limitations due to their environmental sustainability and cost-effectiveness in the long run. It can be hoped that the government and private sectors will form a synergy and work together to overcome some of the policy and financial challenges. If the challenges can be solved by any margin, it will significantly improve the economic condition of developing counties, and above all, a green environment can be assured by doing so.

References

1. N.D. Pham, W. Martens, M.A. Karim, M.U.H. Joardder, Nutritional quality of heat-sensitive food materials in intermittent microwave convective drying. Food Nutr. Res. **62** (2018)
2. N. Duc Pham et al., Quality of plant-based food materials and its prediction during intermittent drying. Crit. Rev. Food Sci. Nutr. **59**(8), 1197–1211 (2019)
3. M.U.H. Joardder, R.J. Brown, C. Kumar, M.A. Karim, Effect of cell wall properties on porosity and shrinkage of dried apple. Int. J. Food Prop. **18**(10), 2327–2337 (2015)
4. M.A. Karim, M.N.A. Hawlader, Mathematical modelling and experimental investigation of tropical fruits drying. Int. J. Heat Mass Transf. **48**(23), 4914–4925 (2005)
5. M.S. Rahman, Dried food properties: Challenges ahead. Dry. Technol. **23**(4), 695–715 (2005)
6. M.U.H. Joardder, M. H. Masud, *A Brief History of Food Preservation BT Food Preservation in Developing Countries: Challenges and Solutions*, M.U.H. Joardder, M. Hasan Masud, Eds., Springer International Publishing, Cham, 57–66 (2019)
7. M.U.H. Joardder, M.A. Karim, Development of a porosity prediction model based on shrinkage velocity and glass transition temperature. Dry. Technol., 1–17 (2019)
8. M.U.H. Joardder, C. Kumar, M.A. Karim, Prediction of porosity of food materials during drying: Current challenges and directions. Crit. Rev. Food Sci. Nutr. **58**(17), 2896–2907 (2018)
9. M.I.H. Khan, R.M. Wellard, S.A. Nagy, M.U.H. Joardder, M.A. Karim, Investigation of bound and free water in plant-based food material using NMR T2 relaxometry. Innov. Food Sci. Emerg. Technol. **38**, 252–261 (2016)
10. M.I.H. Khan, R.M. Wellard, S.A. Nagy, M.U.H. Joardder, M.A. Karim, Experimental investigation of bound and free water transport process during drying of hygroscopic food material. Int. J. Therm. Sci. **117**, 266–273 (2017)
11. M.H. Masud, M.U.H. Joardder, M.A. Karim, Effect of hysteresis phenomena of cellular plant-based food materials on convection drying kinetics. Dry. Technol. **37**(10) (2019)
12. A.S. Mujumdar, *Handbook of Industrial Drying,* CRC press (2006)
13. The Renewable Energy Hub, The Disadvantages of Air Source Heat Pumps (2019), Retrieved from https://www.renewableenergyhub.co.uk/main/heat-pumps-information/the-disadvantages-of-air-source-heat-pumps/, (Accessed on: 25th August, 2019)
14. A.M. Flax, U.S. Patent No. 6,178,762. Washington, DC: U.S. Patent and Trademark Office (2001).
15. M.H. Masud, A.A. Ananno, A.M.E. Arefin, R. Ahamed, P. Das, M.U.H. Joardder, Perspective of biomass energy conversion in Bangladesh. Clean Techn. Environ. Policy **21**(4) (2019)
16. A.R. Nabi, M.H. Masud, Q.M.I. Alam, Purification of TPO (Tire Pyrolytic Oil) and its use in diesel engine. IOSR J. Eng. **4**(3), 1 (2014)

17. M.U.H. Joardder, P.K. Halder, M.A. Rahim, M.H. Masud, Solar pyrolysis: Converting waste into asset using solar energy, in *Clean Energy for Sustainable Development*, Elsevier, 213–235, (2017)

18. E. Houshfar et al., NOx emission reduction by staged combustion in grate combustion of biomass fuels and fuel mixtures. Fuel **98**, 29–40 (2012)

19. C.B.B. Guerreiro, V. Foltescu, F. De Leeuw, Air quality status and trends in Europe. Atmos. Environ. **98**, 376–384 (2014)

20. M. Mourshed, M.H. Masud, F. Rashid, M.U.H. Joardder, Towards the effective plastic waste management in Bangladesh: a review. Environmental Science and Pollution Research, **24**(35), 27021–27046 (2017).

21. European Biomass Industry Association, Challenges related to biomass (2018), Retrieved from http://www.eubia.org/cms/wiki-biomass/biomass-resources/challenges-related-to-bio-mass/, (Accessed on: 25th August, 2019)

22. C. Kumar, M.U.H. Joardder, T.W. Farrell, G.J. Millar, A. Karim, A porous media transport model for apple drying. Biosyst. Eng. **176**, 12–25 (2018)

23. C. Kumar, M.U.H. Joardder, T.W. Farrell, M.A. Karim, Multiphase porous media model for intermittent microwave convective drying (IMCD) of food. Int. J. Therm. Sci. **104**, 304–314 (2016)

24. C. Kumar, M.A. Karim, Microwave-convective drying of food materials: A critical review. Crit. Rev. Food Sci. Nutr. **59**(3), 379–394 (2019)

25. C. Kumar, G.J. Millar, M.A. Karim, Effective diffusivity and evaporative cooling in convective drying of food material. Dry. Technol. **33**(2), 227–237 (2015)

26. M.U.H. Joardder, A. Karim, C. Kumar, Effect of temperature distribution on predicting quality of microwave dehydrated food. J. Mech. Eng. Sci. **5**, 562–568 (2013)

27. A. Collignan, A.-L. Raoult-Wack, A. Thémelin, Energy study of food processing by osmotic dehydration and air drying. Agric. Eng. J. **1**(3), 125–135 (1992)

28. P. P. Lewicki, A. Lenart, Osmotic Dehydration of Fruits and Vegetables. In: Handbook of Industrial drying, 665–681 (1995)

29. A. Lenart, P.P. Lewicki, Energy consumption during osmotic and convective drying of plant tissue. Acta Aliment. Pol. **14**(1) (1988)

30. D. Mastrocola, C.R Lerici, M. Dalla Rosa, Osmotic treatments: Exchange processes and chemical-physical characteristics of the reconstituted products. Food Technology Systems to Produce. **10**(1), 70–75 (1999)

31. P. Lewicki, A. Lenart, Energy consumption during osmo-convection drying of fruits and vegetables. Dry. solids, 354–366 (1992)

Chapter 7
Conclusion

Food waste is one of the most challenging issues of the twenty-first century. Every year more than 1.3 billion tons of vegetables, fruits, dairy, meat, grains, and seafood are being spoiled in the food chain or getting thrown away as municipal solid waste. The market value of the wasted food products was nearly 936 billion USD in 2012, which is higher than the combined GDP of many developing countries (e.g. Bangladesh, Nigeria, Ethiopia, Mali). Moreover, annually, food waste generates 4.4 gigatons equivalent of greenhouse gas (GHG), which is more than the GHG emitted by all the vehicles (e.g. cars, airplanes, ships) on the planet combined. Governments and policymakers must unanimously acknowledge that that food waste is one of the biggest threats to the economy, environment, food safety, and human lives. This book attempted to provide facts and information as well as propose advanced drying systems that can contribute towards ameliorating the crisis of food waste.

This book has critically analyzed the food waste scenario of the world to emphasize the urgency for actions against its further proliferation. Drying has proven itself to be a viable means of preventing food waste. However, conventional drying requires a significant energy resource that is unavailable and/or costly to many developing countries. As a result, foods are being wasted, which can otherwise be used to nourish hungry human beings. Before recommending drying as an effective solution, the book has provided insights into drying by providing detail knowledge about the basic principle of drying and its heat and mass transfer mechanisms.

In order to produce high quality dried food, it is essential to understand the properties of food that appeal to general consumers. An ideal drying technology should be able to maintain quality attributes such as texture, taste, aroma, nutrients, and color. However, in the interest of premium quality dried food, expensive drying techniques are also not desirable. There was a time when consumers were only concerned with extending the shelf life of food through drying. However, as standards of living improved, the consumers are have become more concerned about quality of dried food. Therefore, the principles of drying have expanded beyond the limited mechanical and chemical engineering approach of heat and mass transfer. Chapter 2

M. Hasan Masud et al., *Sustainable Food Drying Techniques in Developing Countries: Prospects and Challenges*, https://doi.org/10.1007/978-3-030-42476-3_7

of this book discussed the various quality aspects of drying, including, physical quality, optical properties, sensory properties, biological properties, and nutritional quality indicators. However, if the selected drying techniques become too costly and energy-intensive, consumers of developing countries would be discouraged to adopt them. As such, discussion of energy, time, and cost requirement was also provided. The method of proposing drying techniques for developing countries is crucial, and it has to consider the economy, policy and industrial progress of diverse demography. Based on the technology needs assessment (TAN) report, the concept of drying technique selection was also included.

Fossil fuel sources are not evenly distributed amongst the globe; therefore, policymakers advocating for sustainable energy technologies. Consumers mostly depend on non-renewable energy to meet their everyday demands. However, the reserve for non-renewable energy is rapidly deteriorating. The world at best can sustain 200 more years with the combined available non-renewable energy resources. Overuse of fossil fuel has accelerated climate change and damaged the eco-system. Hence, the governments and policymakers are transitioning towards renewable energy usage. The available reserve of non-renewable energy sources in developing countries is discussed in Chap. 3. It also highlights the renewable energy potential of the developing countries. The chapter also provides a general discussion on the energy requirement for drying. The potential for integrating renewable energy is also highlighted in this chapter. Since renewable energy adoption is still in the early stages in developing countries, its integration with the drying industry might be met with various barriers. A comprehensive discussion on the barriers to renewable energy integration is provided at the end of this chapter.

Many developing countries already use drying to preserve food in one way or another. The simple operation and easy installation of some conventional drying (e.g., sun drying, biomass drying) technologies are responsible for its ubiquitous practice in various developing countries. Although these dryers are inexpensive, they are mostly inefficient and fail to retain the desired food properties. Despite that, due to financial limitations, consumers of many developing countries still use them on a daily basis. As such, the five most common conventional drying techniques used in developing countries are discussed in Chap. 4. The chapter described both the pros and cons of these conventional drying techniques.

Drying is far from a panacea for the world food wastage problem. Albeit, it can significantly improve the self-life of food, the process requires enormous energy, which increases the cost of dried food. In the wake of a global energy crisis, energy extensive food preservation techniques will become obsolete. Therefore, renewable energy-based drying techniques are the future of the food preservation industry. Chapter 5 of this book discussed improved drying techniques that can operate on renewable energy. Extensive discussion was done on their working principle, efficiency, effectivity, and the produced dried food quality. Each developing country has its unique energy economy and geographical features. Therefore, an in-depth comparison among all the proposed dryers was discussed.

Every thermal system is bound by the laws of thermodynamics; therefore, they are prone to limitations. The proposed drying techniques are also susceptible to certain bounds of applicability. Moreover, renewable energy-based drying techniques have difficult market penetrability. As such, developing countries might face certain challenges while adopting the proposed drying techniques. In order to better prepare them for the barriers of implementation, Chap. 6 extensively discusses the various limitations of proposed drying techniques. Barriers are presented from multi-perspective, including policy, technical, financial and economic angles. It is hoped that the constant progress of science and future research in the field of drying technology would one day solve these challenges. Therefore, by adopting the market available renewable energy-based drying technologies the food waste situation can be improved along with ensuring environmental sustainability.

Printed in the United States
by Baker & Taylor Publisher Services